Progress in Vaccinology

Series Editor: G.P. Talwar, New Delhi, India

Editorial Board:

G.L. Ada, The Johns Hopkins University, Baltimore, USA
F. Horaud, Institut Pasteur, Paris, France
R.V. Petrov, Institute of Immunology, Moscow, USSR
J.B. Robbins, National Institutes of Health, Bethesda, USA
J. Salk, The Salk Institute, San Diego, USA
H. Wigzell, Karolinska Institute, Stockholm, Sweden

Progress in Vaccinology

Volume 1 Contraception Research for Today and the Nineties
Progress in Birth Control Vaccines
G.P. Talwar, Editor

Volume 2 Progress in Vaccinology
G.P. Talwar, Editor

Volume 3 Anti-Idiotypic Vaccines
P.-A. Cazenave, Editor

Progress in Vaccinology
Volume 3

P.-A. Cazenave
Editor

Anti-Idiotypic Vaccines

With 29 Illustrations

Springer-Verlag
New York Berlin Heidelberg London
Paris Tokyo Hong Kong Barcelona

Professor P.-A. Cazenave
28, Rue du Dr. Roux
F-75724 Paris Cedex 15
France

Library of Congress Cataloging-in-Publication Data
Anti-idiotypic vaccines/P.-A. Cazenave, editor.
    p. cm.—(Progress in vaccinology; v. 3)
  Includes bibliographical references.
  ISBN 0-387-97142-4 (alk. paper)
  1. Anti-idiotypic vaccines. 2. Anti-idiotypic antibodies.
  I. Cazenave, Pierre-André. II. Series.
    [DNLM: 1. Antibodies, Anti-Idiotypic—immunology.
  2. Immunotherapy. 3. Vaccines. W1 PR685 v. 3 / QW 805 A629]
  QR189.5.A56A56 1990
  615'. 372—dc20
  DNLM/DLC
  for Library of Congress   90-9485

Printed on acid-free paper.

© 1991 Springer-Verlag New York Inc. Copyright is not claimed for U.S. Government employees.
All rights reserved. This work may not be translated or copied in whole or in part without the written permission of the publisher (Springer-Verlag New York, Inc., 175 Fifth Avenue, New York, NY 10010, USA), except for brief excerpts in connection with reviews or scholarly analysis. Use in connection with any form of information and retrieval, electronic adaptation, computer software, or by similar or dissimilar methodology now known or hereafter developed is forbidden.
The use of general descriptive names, trade names, trademarks, etc., in this publication, even if the former are not especially identified, is not to be taken as a sign that such names, as understood by the Trade Marks and Merchandise Marks Act, may accordingly be used freely by anyone.

Typeset by Thomson Press (India) Limited, New Delhi, India.
Printed and bound by Edwards Brothers, Inc., Ann Arbor, Michigan.
Printed in the United States of America.

9 8 7 6 5 4 2 1

ISBN 0-387-97142-4 Springer-Verlag New York Berlin Heidelberg
ISBN 3-540-97142-4 Springer-Verlag Berlin Heidelberg New York

Series Preface

Vaccines have historically been considered to be the most cost-effective method for preventing communicable diseases. It was a vaccine that enabled global eradication of the dreaded disease smallpox. Mass immunization of children forms the anchor of the strategy of the World Health Organization (WHO) to attain "health for all" status by the year 2000.

Vaccinology is undergoing a dimensional change with the advances that have taken place in immunology and genetic engineering. Vaccines that confer short or inadequate immunity or that have side effects are being replaced by better vaccines. New vaccines are being developed for a variety of maladies. Monoclonal antibodies and T cell clones have been employed to delineate the immunodeterminants on microbes, an approach elegantly complemented by computer graphics and molecular imaging techniques. Possibilities have opened for obtaining hitherto scarce antigens of parasites by the DNA recombinant route. Better appreciation of the idiotypic network has aroused research on anti-idiotypic vaccines. Solid-phase synthesis of peptides is leading to an array of synthetic vaccines, an approach that is expected to attain its full potential once the sequences activating suppressor cells are discovered and the rules for presentation of antigens to T and B cells are better worked out.

A new breed of vaccines is on the horizon that seeks to control fertility. Originally conceived to intercept a step in the reproductive process, they are conceptual models for developing approaches to regulate the body's internal processes. The importance of lymphokines and monokines in the induction of the immune response and in killing parasites is realized, and specific or nonspecific routes are employed to elicit their formation. Interleukins and interferons have been produced by DNA recombinant methods and experimental approaches initiated to coexpress the genes for such regulators with microbial antigens. The old smallpox vaccine, vaccinia, is appearing in a new garb with genetically engineered foreign genes. The technology for manufacture of vaccines, especially for cell culture-based organisms, is undergoing changes, with new cell lines, promoters for better expression, and automation.

Contemporary vaccinology is a multidisciplinary science (and technology) that is developing rapidly. Findings are reported in disparate journals. Periodical reviews by experts assimilating relevant progress in a given field would be of immense value to investigators, funding agencies, manufacturers, and users of the vaccines, the public health authorities. This series aims to provide comprehensive reviews on topics relating to various aspects of vaccinology by leading investigators.

G.P. Talwar

Preface

Numerous vaccination regimes in use to date were developed empirically, in the absence of a thorough understanding of the cellular and molecular basis of immunity. In spite of this, such vaccines have allowed the control, and in some cases the eradication, of infectious diseases.

Traditional vaccines are either attenuated live strains or killed organisms; in some cases the entire organism is substituted by more or less purified microbial antigens. Moreover, classical vaccines can present serious disadvantages. Such a disadvantage can reside in the vaccine itself if live attenuated organisms generate virulent revertants. Alternatively, problems of efficiency can be due to the genetic background and implicitly, to the immune behavior of the recipient (the failure to get protection against tuberculosis by means of vaccination with BCG is well documented for populations of India). Even purified antigenic preparations might not bring optimal results if, for example, one cannot separate toxic from antigenic components via a simple and reliable procedure. On the other hand, the production of synthetic vaccines could also encounter some serious difficulties; in particular if the protective antigen is a carbohydrate, often it cannot be synthesized at an industrial scale. Finally, some pathogens are able to change their antigenic constellation while in the host via a programmed "switching" (a phenomenon well documented in parasites), thus beating immunologists at their own game. Obviously, new and more sophisticated ways of achieving a desired state of immunity must be investigated.

The network hypothesis elaborated by Niels Jerne introduced the concept of internal image and stimulated the idea of manipulating the immune system by means of anti-idiotypic antibodies. It was proposed that internal images (the consequence of mimicry of antigens by anti-idiotypic antibodies) could be used as immunogens for vaccinations.

The purpose of this book is to serve the interested in the practical aspects of idiotypy for vaccines. The first two chapters discuss concepts behind idiotype vaccines. M. Brait and colleagues explain why these could be worthwhile, while R. Bowen and C. Bona analyze what is known about in vivo stimulation of lymphocytes by anti-idiotypic antibodies. The third chapter by C. Williams and colleagues presents the genetic polymorphism of human VH genes families, which is worth considering due to its direct relevance to idiotype vaccines. Several following chapters focus on animal models for specific idiotype vaccines. K. Stein summarizes the situation of carbohydrate antigens and G. Viale and A. Siccardi as well as H. Köhler and S. Müller present data on idiotypic reagents as tumor vaccines. R.Q. Warren and R.C. Kennedy and D.B. Weiner and colleagues address the problem of anti-HIV vaccination. The report of J.-M. Grzych and collaborators deals with the use of anti-idiotypic antibodies as immunoprophylaxis against parasites. In the last contribution M. Zanetti and his colleagues describe recombinant antibody molecules bearing engineered idiotopes.

Pierre-André Cazenave

Contents

Series Preface v
Preface vii
Contributors xi

1. Some Thoughts on the Future of Idiotypic Vaccines   1
   M. Brait, J. Tassignon, J. Ismaili, J. Marvel, K. Meek, O. Leo, and
   J. Urbain

2. Stimulation of Lymphocytic Clones by Anti-Idiotypic Antibodies:
   Basis for Development of Idiotypic Vaccines   8
   Richard Bowen and Constantin Bona

3. Small Human V_H Gene Families Show Remarkably Little
   Polymorphism 22
   Carol Williams, Linda Weigel, Inaki Sanz, and J. Donald Capra

4. Anti-Idiotypic Vaccines for Carbohydrate Antigens 31
   Kathryn E. Stein

5. Characterization of Idiotypic Reagents as Antigen Surrogates of a
   Human Tumor-Associated Antigen 42
   Giovanna Viale and Antonio G. Siccardi

6. Anti-Idiotypic Tumor Vaccines. 55
   Heinz Köhler and Sybille Müller

7. Anti-Idiotypic Antibodies as Potential Viral Vaccines 73
   Ronald Q. Warren and Ronald C. Kennedy

8. Utilization of Anti-Idiotypic Antibodies as Molecular Probes of Virus–Receptor Interaction 92
   David B. Weiner, Daniel E. McCallus, William V. Williams, and Mark I. Greene

9. Anti-Idiotypic Antibodies: An Alternative Approach to Immunoprophylaxis Against Parasites 107
   Jean-Marie Grzych, Florence Velge-Roussel, and André Capron

10. Idiotype Vaccines by Antibody Engineering: Structural and Functional Considerations 123
    Maurizio Zanetti, Rosario Billetta, and Maurizio Sollazzo

Index 138

Contributors

Rosario Billetta, Department of Medicine, University of California at San Diego, San Diego, California, USA
Constantin Bona, Department of Microbiology, Mount Sinai School of Medicine, New York, New York, USA
Richard Bowen, Department of Microbiology, Mount Sinai School of Medicine, New York, New York, USA
M. Brait, The Laboratoire de Physiologie Animale, Université Libre de Bruxelles, Faculté des Sciences, Bruxelles, Belgium
J. Donald Capra, Department of Microbiology, The University of Texas Southwestern Medical Center at Dallas, Dallas, Texas, USA
André Capron, Centre d'Immunologie et de Biologie Parasitaire, Unité Mixte INSERM U167-CNRS 624, Institut Pasteur, Lille Cédex, France
Mark I. Greene, Department of Pathology, University of Pennsylvania, Philadelphia, Pennsylvania, USA
Jean-Marie Grzych, Centre d'Immunologie et de Biologie Parasitaire, Unité Mixte INSERM U167-CNRS 624, Institut Pasteur, Lille Cédex, France
J. Ismaili, The Laboratoire de Physiologie Animale, Université Libre de Bruxelles, Faculté des Sciences, Bruxelles, Belgium
Ronald C. Kennedy, Department of Virology and Immunology, Southwest Foundation for Biomedical Research, San Antonio, Texas, USA
Heinz Köhler, IDEC Pharmaceuticals Corporation, La Jolla, California, USA
O. Leo, The Laboratoire de Physiologie Animale, Université Libre de Bruxelles, Faculté des Sciences, Bruxelles, Belgium
J. Marvel, The Laboratoire de Physiologie Animale, Université Libre de Bruxelles, Faculté des Sciences, Bruxelles, Belgium
Daniel E. McCallus, Department of Pathology, University of Pennsylvania, The Wistar Institute, Philadelphia, Pennsylvania, USA
K. Meek, Department of Microbiology, The University of Texas Health Science Center at Dallas, Dallas, Texas, USA
Sybille Müller, IDEC Pharmaceuticals Corporation, La Jolla, California, USA

Inaki Sanz, Department of Medicine, Division of Rheumatology, The University of Texas Health Science Center at San Antonio, San Antonio, Texas, USA

Antonio G. Siccardi, Dipartimento di Biologia e Genetica per le Scienze Mediche, Universitá di Milano, Milano, Italy

Maurizio Sollazzo, Department of Medicine, University of California at San Diego, San Diego, California, USA

Kathryn E. Stein, Division of Bacterial Products, Office of Biologics Research, Center for Biologics Evaluation and Research, Food and Drug Administration, Bethesda, Maryland, USA

J. Tassignon, The Laboratoire de Physiologie Animale, Université Libre de Bruxelles, Faculté des Sciences, Bruxelles, Belgium

J. Urbain, The Laboratoire de Physiologie Animale, Université Libre de Bruxelles, Faculté des Sciences, Bruxelles, Belgium

Florence Velge-Roussel, Centre d'Immunologie et de Biologie Parasitaire, Unité Mixte INSERM U167-CNRS 624, Institut Pasteur, Lille Cédex, France

Giovanna Viale, Dipartimento di Biologia e Genetica per le Scienze Mediche, Universitá di Milano, Minalo, Italy

Ronald Q. Warren, Department of Virology and Immunology, Southwest Foundation for Biomedical Research, San Antonio, Texas, USA

Linda Weigel, Department of Microbiology, The University of Texas Southwestern Medical Center at Dallas, Dallas, Texas, USA

David B. Weiner, Departments of Pathology and Medicine, University of Pennsylvania, The Wistar Institute, Philadelphia, Pennsylvania, USA

Carol Williams, Department of Microbiology, The University of Texas Southwestern Medical Center at Dallas, Dallas, Texas, USA

William V. Williams, Department of Medicine, University of Pennsylvania, Philadelphia, Pennsylvania, USA

Maurizio Zanetti, Department of Medicine, University of California at San Diego, San Diego, California, USA

CHAPTER 1

Some Thoughts on the Future of Idiotypic Vaccines

M. Brait, J. Tassignon, J. Ismaili, J. Marvel,
K. Meek, O. Leo, and J. Urbain

We would like to answer the questions: Are idiotypic vaccines worthwhile? What do we know? What must we ignore? What should be done in the near future? However, we have not reviewed the work on idiotypic cascades here, since several early and recent reviews or books (including this one) are available (1–4).

In these days and for many immunologists, the situation looks rather pessimistic. Some people even say: "Let us forget about idiotypic vaccines." It is true that in many cases commercial laboratories, which are the only ones with the staff and the money, do not want to invest in idiotypic vaccination (vaccines of the third generation).

It is our point of view that the whole situation should be reconsidered in the light of recent findings, and it is also our point of view that such vaccines can be highly useful in the fight against cancer, AIDS, and certain bacterial diseases of children.

We shall first summarize what we know.

It has been known for some time that idiotypes or anti-idiotypes can profoundly influence immune responses (5, 6). For example in the case of recurrent idiotypes (the animal is programmed in some way to make a given idiotype), minute amounts of anti-idiotypic (anti-Id) antibodies can strikingly enhance the subsequent immune response against the antigen. Higher doses provoke suppression. Doses in between do not change the response but in a very interesting paper, G. Kelsoe has shown that this lack of change in the immune system is in fact due to a balance between activation and suppression (7, 8). We still do not understand dose effects; we do not know why it is necessary to inject the antigen several weeks after anti-idiotypic treatment. The elegant statement of K. Rajewsky, "The authors conclude that quite in the sense of the original network concept, the anti-idiotype disturbs the idiotypic balance of the system and a new equilibrium is reached by a complex series of reactions which proceed over many weeks and in which large numbers of idiotypes, anti-idiotype molecules and cells are involved," is just an elegant way of stressing our ignorance about networks. Who will bring the network into the attic light?

More work should be done to understand these dose effects not only for fundamental immunology but also for applied immunology.

It has been shown, and we were the first, together with the editor of this book P.A. Cazenave (9, 10), to show that injection of polyclonal Ab_2 directed against specific idiotypes or idiotypes "à la Oudin" (ones that are lot-drawn within a large array of possible idiotypes) induce the synthesis of anti–anti-idiotypic antibodies (Ab_3), some of which are very similar to the starting Ab1.

In a formally manner (11), we can expect at least three subsets of antibodies.

1. Idiotype-negative and antigen-nonbinding antibodies: Id − Ag −. These are just antibodies against antibodies.
2. Idiotype-positive and antigen nonbinding antibodies: Id + and Ag −. Experimentally this subset is the most important one in Ab3.
3. Idiotype-positive and antigen-binding antibodies: Id + and Ag +. We have called this subset Ab1'.

Subset 3 is often present in very small amounts. Table 1.1 clearly illustrates this point in the arsonate system.

Table 1.1. Average concentration (μg/ml) of serum antibodies bearing different CRI_A idiotopes in untreated or manipulated Balb/C injected with anti-Id antibodies.

Treatment	ARS–KLH	Anti-ARS	E4 +	E3 +	H8 +	2D3 +	CRI
NR Igs	−	—	—	—	—	—	—
NR Igs	+	260	45	20	20	—	10
RA CRI	−	3	247	64	56	—	347
RA CRI	+	235	1638	690	660	—	1061
−	+	134	22	22	10	—	ND
MAb2 E4	−	—	273	8	43	—	19
MAb2 E4	+	161	545	69	45	—	133
MAb2 E3	−	—	—	37	—	—	—
MAb2 E3	+	147	—	45	—	—	—
MAb2 H8	−	—	138	37	1240	—	17
MAb2 H8	+	120	610	50	560	—	139
MAb2 2D3	−	—	—	—	—	38	—
MAb2 2D3	+	346	—	—	—	67	—
MA2 E4, E3, H8, 2D3	−	—	139	109	116	—	105
MAb2 E4, E3, H8, 2D3	+	12	2388	1203	1255	—	ND
MAb2 F6.51	−	—	—	—	—	—	—
MAb2 F6.51	+	290	13	15	—	—	—

E_3, E_4, H_8, and $2D_3$ are distinct idiotopes of the CRI_A. The binding of peroxydase-labeled, affinity-purified 3665 CRI_A MAb on any of the 4 MAb2(E_4, E_3, H_8, $2D_3$) or on rabbit anti-CRI was inhibited by serial dilutions of the various sera in a standard competition ELISA assay. The titer of anti-ARS Ig is determined by a standard binding ELISA assay in ARS–BSA. Titers (μg/ml) of idiotype bearing Igs and anti-ARS Igs in the sera were extrapolated at 50% inhibition or binding using a standard inhibition curve with unlabeled 3665 MAb. The numbers shown are the geometric average titer in the sera of 20 micees; MAb FG.51, an anti-ID antibody directed against MOPC 460; ND, not done; —, level of positive antibodies under the detection limit of the assay. These limits were ArS, 1 μg/ml; MAb2, 5 to 10 μg/ml.

After anti-idiotypic treatment, injection of the antigen directed against the Ab1 used to induce the vaccine amplifies the Ab1' subset. This has now been done in a very large number of antigenic or idiotypic systems.

However, there is some limitation due to genetic polymorphism. For example, we have shown that Balb/c mice (IgHa) can produce an idiotype that is, idiotypically, strongly cross-reactive with the recurrent idiotype of A/J mice (IgHe), the CRI_A idiotype, despite the fact that the heavy chains are very different (except in the D segment) (11). Balb/c mice and A/J mice have the same IgK locus. We have also shown that C.C58 mice, which have different heavy and light chains (IgHa locus, IgK from C58 mice), can make anti-arsonate antibodies that are strongly cross-reactive with the A/J idiotype(13).

We can change the heavy chain locus to the light chain and the idiotypic cascade can work, but this does not occur in all strains of mice (see Fig. 1.1). Therefore although the genetic polymorphism of immunoglobulin genes is not an absolute barrier, it cannot be ignored in the design of idiotypic vaccines. This is the first limitation of idiotypic vaccines and we shall further discuss the problem later.

We come now to the most serious problem of idiotypic vaccines. It is obvious that an ideal idiotypic vaccine should be a monoclonal Ab2 or monoclonal Ab2 mixed with an efficient, nontoxic adjuvant.

However, it is the experience of several laboratories (including our own) that only a small subset of monoclonal Ab2 can replace polyclonal Ab2 (it seems to be something like 1 out of 10). This is dramatically illustrated in Table 1.1, in which we show that none of our monoclonal Ab2 in the arsonate system is as efficient as the rabbit polyclonal Ab2 serum.

S. Epstein (14) has shown that only one MAb out of seven can be used to induce Ab1' antibodies in the anti-Ia.7 system. Not the idiotype, the fine specificity, or affinity can be used to predict (before the experiment) which MAb has a good internal image.

Is this a "butterfly effect" (extreme sensitivity to initial conditions) (15)?

It has been said that the flight of a butterfly in La Jolla can produce some climatic changes in Brussels, but we know also that we will never be able to forecast correctly (with a confidence of 100%) the weather tomorrow in Paris or in Basel. That is what is told to us by the experts in chaos science, and we expect some chaotic events in such complex networks as the immune system.

We know that the response to Ab2 is strongly T cell dependent. This is shown in Figure 1.1. Nude Balb/c mice are unable to respond to the Ab2.

The new insights into T cells (MHC restriction, processing, agretopes,...) have led us to propose that "good" Ab2 could be the antibody that can stimulate antigen-specific T cells. It is expected that only a few monoclonal Ab2 will be able to do this. By contrast polyclonal Ab2 have a higher chance of activating antigen-reactive T cells. There is clearly a lack of knowledge of the role of T cells in the idiotypic cascade, although it is clear that some Ab2 can induce T cells that recognize the starting antigen.

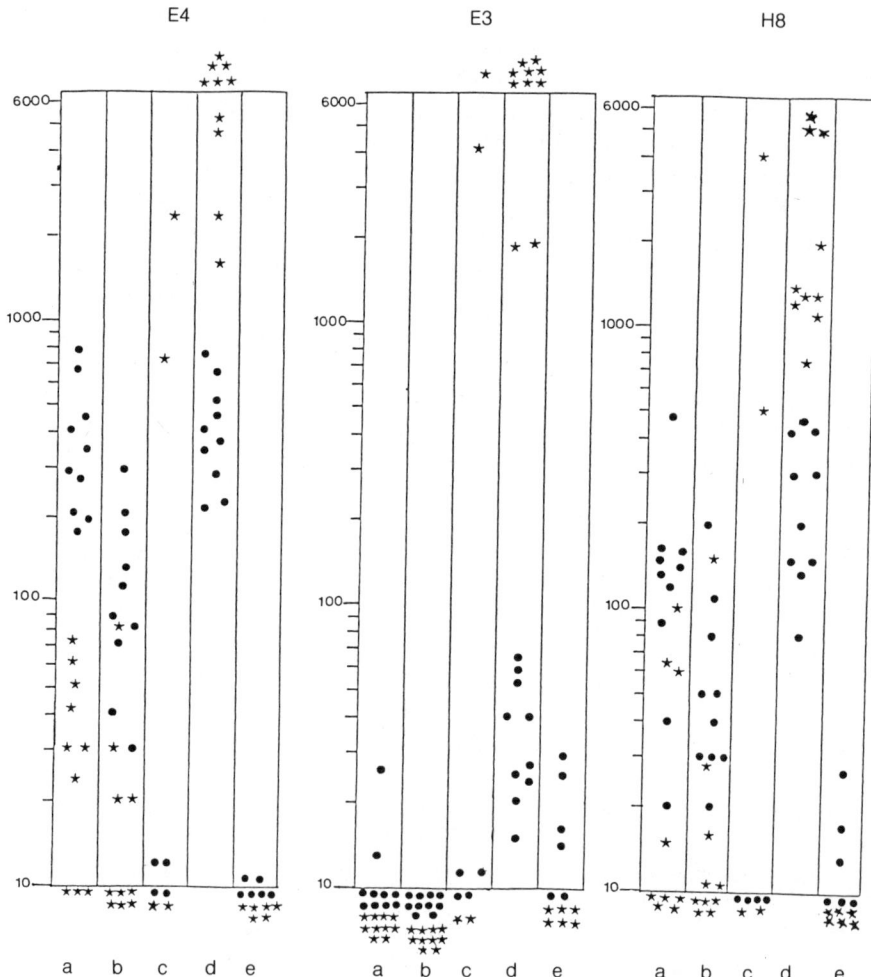

Fig. 1.1. The titers (μg/ml) of idiotype-positive antibodies were defined as described in Table 1.1. The black circles represent mice inoculated with polyclonal rabbit Ab2. We have not detected significant amounts of anti-ArS Ig in these sera. The stars represent mice inoculated with polyclonal rabbit Ab2 and boosted with ARS–KLH. The level of anti-ArS Ig is over 500 μg/ml. At any stage, $2D_3 + $ Ig are not detected. Key to abbreviations: a, AKR; b, C58; c, C. C58; d, Balb/c; e, Balb/c nude.

If the above is correct, we could perhaps bypass this problem by providing good T cell help. We could link an Ab2 with a carrier and then recover the antigen coupled to the same carrier.

However we should not forget another paradox of the immune system: Carrier priming induces hapten-specific suppression. This phenomenon has been recently analyzed by Leclerc's group (16) to show that the phenomenon probably stems from B cell dominance. This phenomenon should be investigated further.

There is perhaps another way to activate good T cell help, thanks to the properties of dendritic cells (DC), which are the most powerful stimulations of T cells ever discovered. A few examples illustrate this point:

1. In vitro primary T cells give an extremely weak response to soluble proteins. However, DC loaded with an antigen can induce a primary response (17).
2. Dendritic cells can transform nonresponders into responders (18).
3. Dendritic cells from Balb/c mice loaded with a polyclonal rabbit Ab2 (an anti–anti-TMV) can induce the synthesis of anti-TMV antibodies in mice that have never seen the virus. Furthermore, a good memory is raised, since the mice treated with DC and injected once thereafter with the virus synthesize 300 μg/ml of anti-TMV antibodies (instead of the 10 μg/ml synthesized by the controls) (19).

We think that syngeneic DC might be replaced with allogeneic DC. A cluster could be formed with allogeneic DC, alloreactive helper T cells, and B cells recognizing the antigen. The alloreactive T cells could furnish the necessary progression factors to the B lymphocytes that are in close proximity. This may possibly solve the MHC-restriction problem in the general use of DC. We can even envision the use of human DC transfected with many HLA genes. In the near future, we should learn more and more about these "magic" tools.

The most obvious question is Where is the power of the DC? Are these cells efficient because they can form clusters and effect contact between relevant lymphocytes? Is it a question of the sialic acids on MHC? Do the cells secrete a specific, unknown lymphokine?

On the basis of these ideas, we can propose a global strategy to obtain an efficient idiotypic vaccine. The first general barrier for all types of vaccine is the genetic polymorphism of immunoglobulin loci MHC.... This barrier can very often be overcome, as shown above. Nevertheless, it could be useful to probe the preimmune repertoire of hundreds of individuals with polyclonal activators to see if the relevant idiotype is present.

The second step is to obtain a large collection of monoclonal Ab2 and to determine which one is able to activate antigen-specific T cells. This should be possible, since DC loaded with an antigen can induce in vitro primary responses. Therefore it is possible to activate T cells against one Ab2 and to test these T cells for proliferation with the starting antigen. One can also imagine that DC are pulsed with the mixture of Ab2 and tested by the same method, provided that the processing requirements of the DC are known (denaturation of the antigen, cleavage, addition of one agretope,...). The reader can easily imagine the subsequent steps.

Finally we must ask the question: What is the purpose of the idiotypic vaccines?

The problem has been discussed many times and most elegantly by K. Eichmann (3). He wrote:

> For most diseases we find it hard to envisage that idiotype vaccines have a realistic chance for competition. However, niches may exist where anti-idiotypic antibodies alone or in conjunction with nominal antigen may facilitate the rational design of a new generation of vaccines. Antiidiotypic antibodies that mimic a protective epitope can be produced even if one does not know the molecular nature of the epitope. Furthermore, mass production of monoclonal antibodies may be less costly and time consuming than the synthesis of certain antigens, e.g. carbohydrates. We see the major field of potential application in individuals who are non or low responders to an antigen and in diseases in which there is a profound dysregulation of the immune system itself.

For these reasons we think that idiotypic vaccines can be useful in tumor immunology and in the treatment of AIDS patients, infants unable to generate an anti-carbohydrate response to *Hemophilus influenzae*, *Neisseria meningitidis*, *Streptococcus pneumoniae*, and also for patients infected with complex parasites.

We hope we will be able to avoid or to control the butterfly effect.

Acknowledgments. This work has been supported by ARC (Belgian State), by the Ministère des Technologies Nouvelles, by I.R.S.I.A., and by F.N.R.S. We are indebted to A. Tassin for expert secretarial help.

References

1. Köhler H, Cazenave PA, Urbain J, eds: *Idiotypy in Biology and Medicine.* Acad Press; 1984.
2. Möller G: Antiidiotypic antibodies as immunogens. *Immunol Rev* 1986; 90.
3. Eichmann K, Emmrich F, Kaufmann S: Idiotypic vaccinations: consideration towards a practical application. *Crit Rev in Immunol* 1987; 7(3):193.
4. Cazenave PA, ed: *Anti-Idiotypic Vaccines.* New York: Springer-Verlag; 1990.
5. Eichmann K, Rajewsky K: Induction of T and B cell immunity by antiidiotypic antibody. *Eur J Immunol* 1975; 5:661–666.
6. Trenkner E, Riblet R: Induction of antiphosphorylcholine antibody formation by antiidiotypic antibodies. *J Exp Med* 1975; 142:1121.
7. Kelsoe G: Regulation of the immune response. II. Concomitant idiotope specific enhancement and suppression can result in a phenotypically normal response. *Cell Immunol* 1986; 98:145.
8. Kelsoe G, Reth M, Rajewsky K: Control of idiotope expression by monoclonal antiidiotope antibodies. *Immunol Rev* 1980; 52:75.
9. Cazenave, PA: Idiotypic-antiidiotypic regulation of antibody synthesis in rabbits. *Proc Natl Acad Sci* 1977; 74:5122.
10. Urbain J, Wikler M, Franssen JD, Collignon C: Idiotypic regulation of the immune system by the induction of antibodies against antiidiotypic antibodies. *Proc Natl Acad Sci* 1977; 74:5126.

11. Antigen on the inside. (1982) ed. by I. Schnurr editions "Roche". p122.
12. Meek K, Jeske D, Slaoui M, Leo O, Urbain J, Capra JD: Complete amino acid sequence of heavy chain variable regions derived from two monoclonal anti-p-azophenylarsonate antibodies of Balb/c mice expressing the major crossreactive idiotype of A/J strain. *J Exp Med* 1984; 160:1070.
13. Marvel J, Tassignon J, Brait M, et al: The influence of VK gene polymorphism of the induction of silent idiotypes in the arsonate system. *Mol Immunol* 1987; 24:463.
14. Misplon J, Kindt R, Reeves J, Harvath L, Rubinstein L, Epstein S: Induction of antigen-specific immunity by anti-idiotypic antibodies: Isotype expression in responses and potency of induction by monoclonal anti-idiotypes. In: Osterhaus A, Uyt de Haag F, eds: *Idiotype networks in biology and medicine*. Amsterdam: Excerpta Medica; 1990, 51.
15. Gleick J: *Chaos*. Viking; 1987.
16. Leclerc C, Herzenberg L: Regulation of antibody production by suppressor T cells. *Res in Immunol* 1989; 140:285.
17. Macatonia S, Taylor P, Knight S, Askonas B: Primary stimulation by dendritic cells induces antiviral proliferative and cytotoxic T cell responses in vitro. *J Exp Med* 1989; 169:1255.
18. Boog C, Kast W, Timmers H, Boe J, De Waal L, Melief C: Abolition of specific immune response defect by immunization with dendritic cells. *Nature* 1985; 318:59.
19. Francotte M, Urbain J: Enhancement of antibody response by mouse dendritic cells pulsed with tobacco mosaic virus or with rabbit antiidiotypic antibodies raised against a private rabbit idiotype. *Proc Natl Acad Sci* 1985; 82:8149.

CHAPTER 2

Stimulation of Lymphocytic Clones by Anti-Idiotypic Antibodies: Basis for Development of Idiotypic Vaccines

Richard Bowen and Constantin Bona

Introduction

Jerne's network theory of the immune system is strongly supported by numerous findings that there are lymphocytic clones bearing receptors that recognize antigenic markers expressed on the receptors of other clones (1).

The antigen receptors of B and T lymphocytes bear three different antigenic determinants: isotypes, allotypes, and idiotypes.

The idiotypes, phenotypic markers of variable (V) genes encoding immunoglobulin (Ig) or T cell receptors (TCR), are the single antigenic determinants recognized in an autologous (or a syngeneic) system.

From an operational point of view, clones bearing a given idiotype are designated Ab1, whereas those that recognize the idiotypes are designated Ab2.

Ab2 clones exist in both B cell and T cell compartments. Thus, there are B cells that produce antibodies able to interact with the idiotypes of Ig receptors of B cells or the TCR of T cells. Similarly, there are T cells that recognize the idiotypes of Ig receptors of B cells or the TCR of other T cells (data reviewed in reference 2) (Fig. 2.1). If we conceive that in the steady state there is an equilibrium between Ab1 and Ab2, then this symmetry of existence of Ab2 between T cell and B cell compartments strongly suggests that the idiotypic network really exists and that experiments aimed at disproving its physiological role are not necessary, since it cannot be disproven.

The population of clones designated as Ab2 is heterogeneous, since we can envision at least three major categories of products that interact with the idiotypes of the lymphocytic receptor.

Ab2α consists of a subset of clones expressing receptors able to recognize the combining site- or framework-associated idiotypes borne by the Ig or the TCR of other clones. The product of such clones can also interact with regulatory idiotopes, which are expressed on the receptors of clones specific for different antigens (2).

Ab2β consists of a subset of clones expressing receptors that bear idiotypes that resemble antigens. Through molecular mimicry, the products of such

Fig. 2.1. Interaction of Ab2 with idiotypes of immunoglobulin receptor of B cell or TCR of T cells.

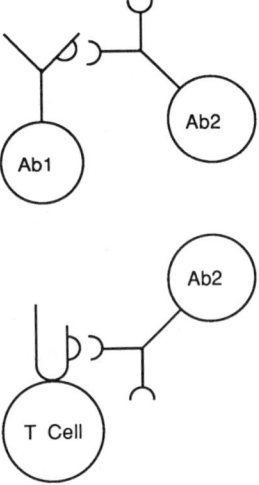

clones can stimulate Ab1 clones in lieu of antigen. The Ab2β carries the internal image of the antigen (3).

Ab2ε consists of a subset of clones expressing receptors that can interact with antigen as well as with idiotypes. The product of such clones are designated *epibodies* (4). Such antibodies play an important role in the clonal connectivity within the immune system (5).

Depending on the concentration of Ab2, they can exert a suppressive or stimulatory effect on clones bearing corresponding idiotypes (data reviewed in reference 2).

In this review we have limited our discussion to the stimulatory effect of anti-idiotypic (anti-Id) antibodies, a topic that is pertinent to our understanding of the development of idiotypic vaccines.

The stimulatory effect of anti-Id antibodies was first reported by Eichman and Rajewsky (6), who showed that the injection of guinea pig anti-A5AId antibodies in A/J mice has a priming effect for an antipolysaccharide response. Later, in rabbits it was shown that the injection of allogenic anti-Id antibodies was able to expand a heterogeneous clonal population composed of Ab3–anti–anti-Id antibodies and a minor fraction producing Ab1 (Ab1′) similar to those elicited by immunization with antigens (7, 8).

The stimulatory effect of syngeneic anti-Id antibodies was first shown in the MOPC460Id system, in which the administration of polyclonal or monoclonal anti-460Id antibodies followed by immunization with trinitrophenol (TNP)–T-independent antigens led to the activation of MOPC460Id+ clones, which represent a minor component of the anti-TNP response (9).

An important aspect of studies of the stimulatory aspect of anti-Id antibodies was provided by the information obtained from the experiments of Hiernaux et al (10), who showed that the injection after birth of minute amounts of anti-Id antibodies could expand silent clones.

The activation of silent clones by anti-Id antibodies was important to the development of the concept of idiotypic vaccines, since it indicated that the anti-idiotype can be used not only to expand the clones against a given microbe but also to expand the silent clones that have the potential to interact with genetic variants of the same microbe. Soon after, Sacks et al (11) showed that anti-Id antibodies could be used to elicit specific immunity against *Trypanosoma*, a parasite exhibiting genetic variation during its growth in host cells.

The major questions concerning the stimulatory effect of anti-Id antibodies are related to (a) the ability to distinguish Ab2α from Ab2β, a matter of practical importance for the selection of anti-Id antibodies that are endowed with Ab2 properties and that can potentially serve as useful vaccines and (b) the requirements of MHC-restricted recognition of anti-Id antibodies.

Several criteria have been proposed to distinguish Ab2α from Ab2β. The most faithful is a structural one represented by amino acid sequences shared by antigen and anti-Id antibody. Indeed, there are two examples that clearly demonstrate this phenomenon. One is that the CDR2 of antibody exhibiting Ab2β properties has a short sequence identical to that found in the protective epitope of the hemagglutinin on reovirus (12). Another is that an anti-Id antibody able to expand the clones specific for the GT (glutamic acid-tyrosine) determinant of the GAT (glutamic acid-alanine-tyrosine) terpolymer has a GTT sequence in the CDR3 of the V_h region (13). It is difficult, however, to use a structural criterion to select Ab2β in every case. This is especially true when an Ab2β mimics antigens of nonprotein origin (e.g., polysaccharides). Therefore two other criteria are generally used to distinguish Ab2β from Ab2α:

The *immunochemical criterion* consist of antigenic competition between Ab2β and Ab1. If the idiotype of the Ab2β mimics the Ag then the antigen should inhibit the binding of Ab2β to Ab1.

In the, *functional criterion*, both Ab2α and Ab2β can have a stimulatory effect. However, the outcome of clonal stimulation by these two types of antibodies may be different, whereas Ab2α should, in principle, stimulate clones bearing corresponding idiotypes that can have different specificities (Ag+Id+, Ag−Id+). The Ab2β by virtue of the ability of idiotype to mimic antigen would stimulate only antigen-specific Ab1, which could be idiotype-positive or -negative (Ag+Id+, Ag+Id−). Furthermore, because of molecular mimicry, in principle, an Ab2β should have not only a priming effect but should also be able, as a nominal antigen, to stimulate antigen-specific lymphocytes to differentiate into plasma cells.

Stimulatory Effect of Ab2α

The binding of various biologically active substances to a multitude of receptors associated with T and B lymphocytes can induce the activation, proliferation, and differentiation of small resting lymphocytes into effector cells. This has been known for a long time for lectins that bind to receptors

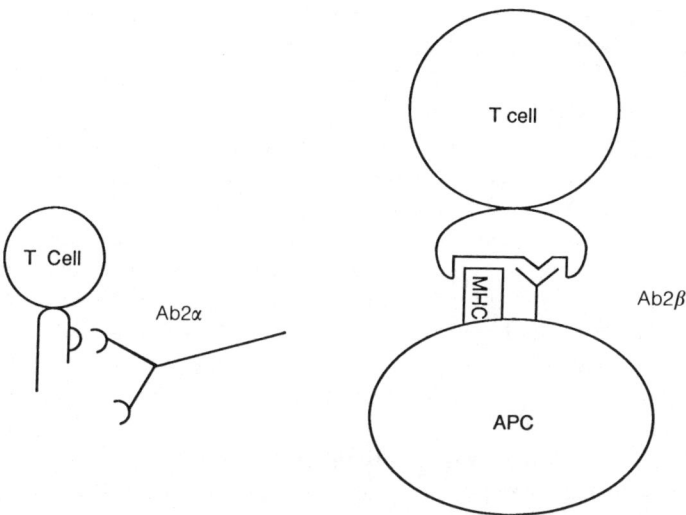

Fig. 2.2. MHC non-restriction or restriction in TCR binding by Ab2α and Ab2β, respectively.

associated with T cell membranes or for polyclonal activators of bacterial origin that bind to receptors associated with B cell membranes. A similar effect was observed with antibodies specific for lymphocytic antigen receptors. Thus it was clearly demonstrated that antibodies specific for allotypic determinants of Ig or TCRs can induce the activation and proliferation of lymphocytes (14–16). Recently it was shown that anti IgM antibody can activate resting small lymphocytes and that the maturation of these activated cells requires lymphokines such as IL-4 secreted by T cells (17). Similarly, anti-idiotypes can activate the clones expressing the corresponding idiotype. In the case of T cells, although Fab fragment suffices to stimulate T cell clones (18), it has not been clearly established that stimulation of B cells by anti-Id antibody requires a single signal resulting from the interaction of anti-idiotype with the idiotype of the B cell receptor or, in addition, one resulting from binding the Fc receptor of B cells. In principle, the activation process by Ab2β of B cell and T cell clones resulting from idiotype–anti-idiotype interaction is not MHC restricted and probably does not require a processing event (Fig. 2.2). The non-MHC restriction is clearly demonstrated by experiments that show that activation of T cells can be achieved with anti-idiotype covalently linked to inert beads. In the case of B cells, we have proposed that anti-Id activation resembles activation induced by T-independent antigens (2).

Stimulation of Humoral Responses by Ab2 α

During the last 5 years, we have studied the stimulatory effect of syngeneic monoclonal anti-Id antibodies in the influenza virus system. We used several syngeneic, anti-Id antibodies interacting with a cross-reactive idiotype expressed on monoclonal antibodies (MAbs) specific for PR8 (H1) and X31 (H3) hemagglutinin as well as a MAb specific for PR8 (N1) and X31 (N2) neuraminidase. The effectiveness of these antibodies as idiotypic vaccines was tested in syngeneic Balb/c mice and allogeneic C3H mice, and their effect was determined by obtaining the hemagglutination inhibition titer of serum antibodies and neuraminidase-specific antibodies by a neuraminidase inhibition assay, as well as by virus titers in lungs. In studies of anti-hemagglutinin antibodies, the animals treated with anti-idiotype were challenged by intraperitoneal injection of influenza virus or by aerosol infection. Multiple parameters were studied to determine the effect of syngeneic anti-Id antibody on the anti-hemagglutinin antibody response: (a) priming of the animal with anti-idiotype in saline, alum precipitate, or linked to keyhole limpet hemocyanin (KLH). (b) doses ranging from 100 ng to 50 µg per mouse; (c) challenging animals with optimal or suboptimal virus doses; and (d) varying intervals between priming with anti-idiotype and virus challenge.

After evaluation of the various parameters, we adopted a scheme of immunization that consisted of priming with 5 µg alum-precipitated anti-idiotype on day 0, repeat priming on day 21 with 5 µg anti-idiotype in saline, and challenge on day 42 with 1 µg virus intraperitoneally or 2.5 µg virus by aerosol infection. The data presented in Table 2.1 show the hemagglutinin inhibition (HI) titers and idiotype levels in animals preimmunized with anti-Id SP3-5A. As can be seen, a 2 log 2 units increase above animals primed with saline was observed on day 21 after challenge with PR8. The increase in HI titer paralleled an increase in serum idiotype.

In spite of the increase in antibody titer, we did not observe differences in lung virus titer in animals primed with anti-idiotype versus control animals. These data suggest that the syngeneic anti-idiotypes used in this study were of the Ab2α type, able to prime the animals but not able to elicit an antiviral immune response by themselves (19).

Similar results have been obtained with syngeneic anti-Id antibodies recognizing cross-reactive idiotypes (CRI) on N1- and N2-specific MAb. A significant anti-NA response was observed in mice injected with 100 ng of anti-idiotype and boosted with PR8, as compared with those injected with MAbs specific for an irrelevant idiotype. However, no anti-NA antibody was observed in the animals primed and boosted with anti-idiotype (20). This result clearly indicates that this particular anti-idiotype behaves as an Ab2α able to prime the cells. Our results taken together show the difficulties in obtaining Ab2β in a system in which the antibody repertoire against complex protein antigens like HA and NA of influenza viruses is so large. This repertoire has about 1500 paratypes (21).

Another example in which Ab2α antibodies can prime a humoral response but are unable to effect B cell maturation in the absence of nominal antigen involves IgM-mediated enhancement of an anti-sheep red blood cell (SRBC), response in mice (22). In this system, an intravenous injection of either purified

Table 2.1. Hemagglutinin inhibition titers and idiotype levels in animals preimmunized with anti-idiotype. With permission from *Vaccines: New Concepts and Developments*, T. Moran, R. Mayer, and C. Bona, copyright © 1987, John Wiley & Sons, Inc.

Immunization[a]			Serum titers	Day 0	Day 3	Day 7	Day 21
1°	2°	Challenge					
Saline	Saline	PR8	HI[b]	0	3 + 0	6 + 0	5 ± 0.63
			IdX[c]	<1	8.72 ± 2.36	1.84 ± 3.68	0.2 ± 0.4
PR8	Saline	PR8	HI	4.8 ± 1.46	4.2 ± 1.16	7.4 ± 1.01	7 ± 0.89
			IdX	<1	8.22 ± 5.6	1.6 ± 1.4	0.9 ± 1.8
SP35A(5 μg)	SP35A	PR8	HI	0 + 0	2.8 ± 0.4	6.8 ± 0.74	7 ± 0.89
			IdX	<1	3.74 ± 4.2	7.68 ± 9.5	25.14 ± 9.6
SP35A(5 μg)	SP35A	Saline	HI	1.2 ± 0.74	1.2 ± 0.75	0	1.8 ± 2.4
			IdX		<1	ND	ND
Saline	Saline	X-31	HI	0	0.4 ± 0.49	5 ± 0	5 ± 0.63
			IdX	<1	<1	7.06 ± 3.6	9.24 ± 6.29
X-31	Saline	X-31	HI	3.6 ± 0.49	4 + 0	5.8 ± 0.75	7.6 ± 0.8
			IdX	<1	<1	23.84 ± 4.7	12.2 ± 4.1
SP35A(5 μg)	SP35A(5 μg)	X-31	HI	0	0.2 ± 0.4	5.5 ± 0.5	5.75 ± 0.43
			IdX	<1	<1	21.54 ± 5.1	7.74 ± 2.7
SP35A(5 μg)	SP35A(5 μg)	Saline	HI	0	0	0	0
			IdX	<1	<1	ND	ND

[a] Animals were immunized with anti-idiotype, saline or virus on day −42(1°) or day −21(2°) and challenged with the indicated virus.
[b] Hemagglutinin inhibition in log₂ units.
[c] Percentage inhibition of PY206 bindings to SP3-5A by a 1:10 dilution of serum.

serum IgM or monoclonal IgM specific for SRBCs were able to enhance the plaque-forming cell (PFC) response in CBA/H mice. These authors also demonstrated that a significant anti-SRBC PFC response could not be induced by injection of IgM alone.

There are two exceptions to the notion that an Ab2α is unable to induce B cell proliferation and antibody formation in the absence of nominal antigen. In one case, rabbit anti-Id antibodies raised against rabbit anti-tobacco mosaic virus (TMV) antibodies and, when coupled to LPS, (lipopolysaccharide) could induce anti-TMV antibodies in Balb/c mice that had never seen the antigen (23). In another case, anti-Id antibodies to a mouse MAb specific for hepatitis B surface antigen were produced in rabbits. These anti-idiotypes, when coupled to KLH and injected into Balb/c mice, induced an anti-HBsAg response without subsequent exposure to antigen (24).

It should be noted in each of these exceptions that the anti-Id reagent used was polyclonal. For this reason, one cannot rule out the possibility that a small proportion of the anti-Id antibodies are Ab2β (internal image) and do not require nominal antigen to elicit an antibody response.

Stimulation of Cell-Mediated Immune Responses by AB2α

The binding of anti-Id antibody can activate and stimulate the proliferation of T cell clones. This phenomenon is well documented for antibodies specific for antigenic determinants encoded by certain Vbeta families. The activation pattern is comparable to that induced by antibody specific for CD3, which is associated with the TCR. Recently, by using hybrid–hybridoma fusion techniques we succeeded in selecting a clone resulting from the fusion of a hybridoma producing anti-X31 HA antibodies (PY206) with a hybridoma producing antibodies specific for the Vbeta8 idiotype of the TCR (F23.1). One of these clones obtained from the fusion (HH8) produced a hybrid bifunctional molecule made up of an Fab fragment and constant region (gamma2b) of PY206 in one half, and an Fab fragment and constant region (gamma2a) of F23.1 in the other half. This bifunctional antibody was able to activate cytotoxic T lymphocyte (CTL) clones and to focus their lytic activity on a melanoma cell infected with X31 influenza. These results clearly show that the binding of an anti-Id antibody of the alpha type, through its Fab, is able to activate (CTLs).

Similar results have been obtained by other investigators. In a study by Ertl et al (25), a Sendai virus-specific T helper cell clone, was used to immunize B10.D2 mice for the production of monoclonal anti-Id antibodies. One hybridoma (1B4.E6) produced an IgM that bound specifically to the clone used for immunization but failed to bind T cells from unimmunized mice or T cells directed against an unrelated antigen. To determine whether anti-Id antibodies could induce a T cell response in an adoptive transfer system, 1B4.E6 culture supernatants were injected intravenously into naive recipient mice. When these mice were challenged in their left footpad with Sendai virus, a local delayed type hypersensitivity (DTH) reaction resulted. This response was antigen specific, T cell dependent, and unlike Sendai virus immunization, MHC unrestricted, since splenocytes from 1B4.E6 immunized donor mice (B10.D2, H-2^d) could induce a DTH reaction in C57B1/6 (H-2^b) recipients and vice versa. When mice were immunized in vivo with this anti-Id antibody, and splenocytes were subsequently restimulated in vitro with virus, they were able to kill virally infected targets in a chromium release assay. Again, this effect was MHC unrestricted. When the antibody was tested in an in vivo protection experiment, anti-Id-primed mice had more than a 10^4-fold reduction in the number of infectious virions in their lungs compared to unimmunized controls.

Another example of the use of anti-idiotypic antibodies against T cell epitopes comes from the work of Kaufmann et al (26), who developed a T cell hybridoma specific for the intracellular bacterial pathogen *Listeria monocytogenes*. This hybridoma was made by fusing a T cell clone conferring specific protection against this bacterium with the AKR thymoma line BW5147. Antisera raised against this hybridoma (TLm1), which specifically blocked IL-2 secretion after its stimulation by heat-killed *Listeria*, was used in vaccination experiments. These antisera conferred protection to mice against *Listeria* infection as measured by the number of bacteria in the spleens of intravenously infected mice. Protection was enhanced by the use of complete Freund's adjuvant, confined to the *Listeria* strain to which TLm1 responded and, as expected for an Ab2α, MHC unrestricted.

Stimulatory Effect of Ab2β

The concept of internal image arises as a statistical necessity of the network theory from the idea that since each paratope of an Ig molecule recognizes the idiotope of a lymphocyte receptor, foreign epitopes must therefore cross-react with idiotopes. Studies carried out by Sege et al (27) showed that, indeed, anti-Id antibodies are the best candidates for representing the internal image. In this experiment, anti-Id antibodies produced against anti-insulin antibodies were able to bind to the insulin cell receptor and to mimic some of the physiological properties of the hormone.

Stimulation of Antibody Responses by Ab2β

In 1983, we showed that a MAb specific for the A48 idiotype shared by antibodies recognizing β2-6 fructosan was able to elicit an anti-levan PFC response in normal mice. Like other anti-Id antibodies from our collection, MAb 17-38 bound to several monoclonal anti-fructosan antibodies. This binding was inhibited by antigen. The injection of this antibody in mice treated at birth with a minute amount of syngeneic polyclonal anti-A48 antibodies led to the activation of clones able to produce anti-levan antibodies. It should be mentioned that the ability of 17-38 to stimulate an anti-levan response was not related to th isotype of Ig, since several antibodies that share the isotype tested for this effect had no Ab2β activity.

In our series of syngeneic monoclonal anti-A48 antibodies, we found only one (17-38) that exhibited Ab2β properties, which indicated that the frequency of Ab2β among anti-idiotypes is very low (28). These data are in agreement with those obtained in the reovirus system. In this system, Gaulton et al (29) also found only one monoclonal anti-Id antibody that carried the internal image of a viral HA protective epitope.

This MAb, an IgM, demonstrated the properties of an Ab2β well. In addition to the primary amino acid sequence matching, this antibody was able to induce anti-reovirus neutralizing antibodies without prior exposure of the animals to nominal antigen. It also induced antiviral antibodies without strain or species restriction. This anti-idiotype gave excellent results when tested for its ability to confer protection in neonatal mice through the immunization of pregnant mothers.

Hepatitis B virus has also been the subject of a series of anti-idiotypic studies. Anti-idiotypic antibodies that exhibit internal image properties have also been produced in this system. These anti-idiotypes react with a cross-reactive idiotope (CRI) on the antibody specific for the surface antigen of hepatitis B. In studies with a polyclonal rabbit anti-idiotype (30), this CRI was detected in sera of most vaccinated or naturally infected individuals. In addition, this idiotope could be detected in anti-hepatitis B surface antigen (anti-HBsAg) sera from several mammalian species but not from chickens. Tests of this rabbit anti-idiotype in an HBsAg-specific plaque-forming cell assay demonstrated an increase in IgM-secreting splenocytes when idiotype was given in saline and an increase in IgG-secreting cells when alum-precipitated Ab2 was used. In a chimpanzee protection study, two chimpanzees that received Ab2 produced detectable anti-HBsAg responses and were completely protected from clinical signs of HBV infection. Although the authors state that this anti-idiotype represented a true internal image, based on the induction of anti-HBsAg in another species with a rabbit Ab2, the nonreactivity of this reagent with chicken anti-HBsAg sera suggests that the Ab2 recognized an idiotype that was closely related but not identical to the paratope of the Ab1 (31).

Studies by Thanavala et al (32) in which MAb2 was utilized also demonstrated the existence of an interspecies CRI from different mammalian species immunized with HBsAg.

Taking a slightly different approach, Colucci and his colleagues (33, 34) used monoclonal anti-idiotypes that mimic polymeric human serum albumin (poly-HSA) and bind to its receptor on HBsAg, to produce syngeneic monoclonal anti–anti-idiotype (Ab3) that could bear the internal image of HBsAg. Some Ab3 appeared to mimic HBsAg, since they reacted with poly-HSA and inhibited monoclonal and polyclonal anti-HBsAg binding to HBsAg. One such Ab3 was injected into rabbits and was able to induce anti-HBsAg antibodies. Thus because it inhibited idiotype–anti-idiotype binding and induced an immune response across species, this anti-idiotype appear to have Ab2β properties.

Another anti-idiotype with Ab2β properties was utilized in studies with poliovirus type II by Uytdehaag and Osterhaus (35). In this study, syngeneic monoclonal anti-idiotype raised against an idiotope on a protective MAb specific for poliovirus. This anti-idiotype was able to induce antibodies with specificity toward the protective epitope but was unable to protect against lethal challenge with poliovirus.

The use of anti-idiotypes in a bacterial system was demonstrated with *Streptococcus pneumoniae* (36). This study utilized the MAb 4C11 that binds an idiotope on an antibody against phosphorylcholine (PC). Such antibodies have been shown to be effective in protecting mice against a lethal streptococcal infection. Two doses of this antibody significantly elevated the serum anti-PC response in Balb/c mice. Antibodies against nominal antigen were produced without exposure to that antigen. The same anti-idiotype treatment gave good protection in mice infected with this bacterium.

An example of the effectiveness of Ab2β in an antiparasite response is given by the work of Grzych et al (37) with *Schistosoma mansoni*. Rat anti-Id antibodies were produced against an Ab1 specific for a schistosomulum membrane target antigen. This antigen is a glycoprotein (mw 38,000), which is strongly immunogenic in humans and several other animal species. The anti-Id antibodies inhibit binding of Ab1 to the target antigen, and serum from rats immunized with purified Ab2 contained target antigen binding Ab3 antibody. These Ab3s contributed to strong cytotoxic reactions against schistosomula and conferred significant protection by passive transfer. Rats immunized with Ab2 were well protected upon schistosoma challenge.

In spite of the numerous data mentioned above that demonstrate that Ab2β are able to stimulate antibody-producing clones, little is known about the cellular requirement and intimate processes of this phenomenon. It would be important to determine whether Ab2β exert their stimulatory effect subsequent to interaction with the Ig receptor or, alternatively, if subsequent to processing, they are recognized in association with class II antigen by T cells that are cooperatively interacting with B cell clones. There are no available data on the stimulatory effect of Ab2β in nude or thymectomized mice.

Stimulation of Cell-Mediated Immune Responses by Ab2β

Numerous data demonstrate that the immune system contains clones able to recognize the idiotype of Ig or of the TCR. It appears that the recognition of idiotypes by T cells follows rules similar to the recognition of foreign antigen, and requires processing and MHC restriction. The requirement for processing was clearly shown in a recent experiment carried out with T cell clones specific for idiotopes borne on the lambda$_2$ chain of MOPC315 (38). Such clones were stimulated in the presence of antigen-presenting cells and the native MOPC315 lambda$_2$ chain. A synthetic peptide corresponding to amino acid residues 91 to 108 was able to stimulate the proliferation of T cell clones in the presence of viable or fixed antigen-presenting cells. It is clearly shown that processing of the native molecule is required for the stimulation of idiotype-specific T cells. The requirement for idiotope processing strongly suggests that the recognition of peptide idiotopes is genetically restricted and under Ir gene control. Waters et al (39), indeed, show that the proliferation of two A48-specific T cell clones upon incubation with irradiated A48 Id-positive hybridoma was inhibited by anti-I-A antibody. Similarly Weiss and Borgen (38) showed that the T cell clones specific for the lambda$_2$ MOPC315 idiotype are inhibited by anti-I-E antibodies. These results taken collectively strongly suggest that the recognition of idiotypes by idiotype-specific T cells is MHC restricted. Therefore, it is predicted that Ab2β mimicking a foreign antigen would be able to stimulate CD4 or CD8 T cell clones in a genetically restricted manner. Thus, for example, an influenza virus-specific CD8 CTL clone able to recognize viral protein with class I MHC antigens can, in principle, be stimulated by Ab2β only by recognizing the idiotype that mimics the viral antigen in association with class I. Similarly, CD4 T cells that recognize the viral antigen in association with class II will be able to be stimulated by the idiotype associated with class II antigen. Therefore, we predict that although the expansion of CD4 or CD8 T cells by Ab2α subsequent to interaction between anti-idiotype and the idiotype of the T cell receptor is not MHC restricted, the activation of T cell clones by Ab2β, through its idiotype, which mimics the antigen, will be MHC restricted.

Few experimental data support our view. Rees et al (40) have described an anti-Id antibody (anti-IdTB71) that contains an internal image in a T cell stimulatory domain that corresponds to a 36-kDa protein antigen from *Mycobacterium tuberculosis*. As already noted, if it is indeed an Ab2β, we would expect this molecule to activate T cells via its interaction with the T cell receptor. The requirement for MHC (HLA-matched accessory cells in this response was examined by determining the ability of HLA-matched and -mismatched accessory cells to support the proliferation of an antigen-specific human T cell clone. Only cells matched at the A and DR locus could present the anti-idiotype to the 38-kDa–specific T cells. Thus, as predicted, the mechanism of presentation of internal image antibody to T cells is similar to the presentation of nominal antigen (i.e., it is MHC restricted).

In closing, we should bear in mind that all of the elements discussed above are important in the selection of Ab2β. In addition, it is clear that the criteria for selection of Ab2β for humoral and cellular immunity are different.

Acknowledgment. This work was supported by NIH grants AI-2446002 and AI-18316.

References

1. Jerne NK: Towards a network theory of the immune response. *Ann Immunol (Paris)* 1974; 125C:373–389.
2. Bona C: *Regulatory Idiotopes*. New York, Wiley 1987.
3. Jerne NK, Roland J, Cazenave PA: Recurrent idiotopes and internal images. *EMBO J* 1982; 1:243–247.
4. Bona C, Finley S, Waters S, Kunkel HG: Anti-immunoglobulin antibodies III. Properties of sequential anti-idiotypic antibodies to heterologous anti-globulins. Detection of reactivity of antiidiotype antibodies with epitopes of Fc fragments (homobodies), and with epitopes and idiotopes (epibodies). *J Exp Med* 1982; 156:986–999.
5. Dwyer DS: Idiotype connectivity of antibody responses specific for self- and nonself antigens. In: Bona CA, ed: *Biological Applications of Anti-Idiotypes, II*. Boca Raton, Fla: CRC Press; 1988; p 55.
6. Eichmann K, Rajewsky K: Induction of T and B cell immunity by antiidiotypic antibody. *Eur J Immol* 1975; 5:661–666.
7. Cazenave P-A: Idiotypic-anti-idiotypic regulation of antibody synthesis in rabbits. *Proc Natl Acad Sci USA* 1977; 74:5122–5125.
8. Urbain J, Wikler M, Franssen JD, Collignon C: Idiotypic regulation of the immune system by the induction of antibodies against antiidiotypic antibodies. *Proc Natl Acad Sci USA* 1977; 74:5126–5130.
9. Bona C, Hooghe R, Cazenave P-A, LeGuern C, Paul WE: Cellular basis of regulation of expression of idiotypes II. Immunity to anti-MOPC460 idiotype antibodies increases the level of anti-trinitrophenyl antibodies bearing 460 idiotypes. *J Exp Med* 1979; 149:815–823.
10. Hiernaux J, Bona C, Baker P: Neonatal treatment with low doses of antiidiotypic antibody leads to the expression of a silent clone. *J Exp Med* 1981; 153:1004–1008.
11. Sacks DL, Esser K, Sher A: Immunization of mice against African trypanosomiasis using antiidiotypic antibodies. *J Exp Med* 1982; 155:1108–1119.
12. Bruck C, Co SM, Slaoui M, et al: Nucleic acid sequence of an internal image-bearing monoclonal anti-idiotype and its comparison to the sequence of the external antigen. *Proc Natl Acad Sci USA* 1986; 83:6578–6582.
13. Fougereau M, Cambillau C, Corbet S, et al: Molecular basis of anti-idiotype antibodies carrying internal image of antigens. In: Bona CA, ed: *Biological Applications of Anti-Idiotypes, I*. Boca Raton, Fla: CRC Press 1988: p 23.
14. Revillard J-P, Brochier J, Vincent C: Modulation of human lymphocyte responses by antibodies directed against lymphocyte surface antigens. In: Bona C, Cazenave P-A, eds: *Lymphocytic Regulation by Antibodies*. New York; Wiley 1981; p 55
15. Infante AJ, Infante PD, Gillis S, Fathman CG: Definition of T-cell idiotopes using antiidiotypic antisera produced by immunization with T-cell clones. *J Exp Med* 1982; 155:1100–1107.
16. Bigler RD, Fisher DE, Wang CY, Kau EAR, Kunkel HG: Idiotype-like molecules on cells of a human T cell leukemia. *J Exp Med* 1983; 158:1000–1005.
17. Paul WE, Ohara J: B-cell stimulatory/Factor-1/interleukin 4. *Ann Rev Immunol* 1987; 5:429–459.
18. Kaye J, Janeway CA: Fab fragment of a directly activating monoclonal antibody that precipitates a disulfide-linked heterodimer from a helper T cell clone blocks activation by either allogeneic Ia or antigen and self-Ia. *J Exp Med* 1984; 159:1397–1412.

19. Moran T, Mayer R, and Bona C: Enhancement of anti-influenza response by anti-idiotype antibodies. In: Kohler H, LoVerde PT, eds: *Vaccines: New Concepts and Developments*. New York, Wiley; 1987: p 342.
20. Mayer R, Ioannides C, Moran T, Johansson B, Bona C: Effect of syngeneic anti-idiotypic antibody on influenza virus neuraminidase antibody response. *Viral Immunol* 1987; 1:121–134.
21. Staudt LM, Gerhard W: Generation of antibody diversity in the immune response of BALB/c mice to influenza virus hemagglutinin. *J Exp Med* 1983; 157:687–704.
22. Heyman B, Andrighetto G, Wigzell H: Antigen-dependent IgM-mediated enhancement of the sheep erythrocyte response in mice: Evidence for induction of B cells with specificities other than that of the injected antibodies. *J Exp Med* 1982; 155:994–1009.
23. Francotte M, Urbain J: Induction of anti-tobacco mosaic virus antibodies in mice by rabbit antiidiotypic antibodies. *J Exp Med* 1984; 160:1485–1494.
24. Schick MR, Dreesman GR, Kennedy RC: Induction of an anti-hepatitis B surface antigen response in mice by noninternal image (Ab_2a) anti-idiotypic antibodies. *J Immunol* 1987; 138:3419–3425.
25. Ertl HC, Finberg RW: Sendai virus specific T cell clones: Induction of cytolytic T cells by an antiidiotypic antibody directed against a helper T cell clone. *Proc Natl Acad Sci USA* 1984; 81:2850–2854.
26. Kaufmann SHE, Eichmann K, Muller I, Wrazel LJ: Vaccination against the intracellular bacterium *Listeria monocytogenes* with a clonotypic antiserum. *J Immunol* 1985; 134:4123–4127.
27. Sege K, Paterson PA: Use of antiidiotypic antibodies as cell-surface receptor probes. *Proc Natl Acad Sci USA* 1978; 75:2443–2447.
28. Rubenstein LJ, Goldberg B, Hiernaux J, Stein KE, Bona CA: Idiotype-antiidiotype regulation V. The requirement for immunization with antigen and monoclonal antiidiotypic antibodies for the activation of B2 6 and B2 1 polyfructosan reactive clones in BALB/c mice treated at birth with minute amounts of anti-A48 idiotype antibodies. *J Exp Med* 1983; 158:1129–1144.
29. Gaulton GN, Sharpe AH, Chang DW, Fields BN, Greene MI: Syngeneic monoclonal internal image anti-idiotypes as prophylactic vaccines. *J Immunol* 1986; 137:2930–2936.
30. Kennedy RC, Ionescu-Matin I, Sanchez Y, Dreesman GR: Detection of interspecies idiotypic cross-reactions associated with antibodies to hepatitis B surface antigen. *Eur J Immunol* 1983; 13:232–235.
31. Kennedy RC, Eichberg JW, Lanford RE, Dreesman GR: Anti-idiotype antibody vaccine for type B viral hepatitis in Chimpanzees. *Science* 1986; 232:220–223.
32. Thanavala YM, Bond A, Tedder R, Hay FC, Roitt IM: Monoclonal internal image anti-idiotypic antibodies of hepatitis B surface antigen. *Immunology* 1985; 55:197–204.
33. Colucci G, Waksal SD: Interactions between hepatitis B virus and polymeric human albumin I. Production of monoclonal anti-idiotypes (anti–anti-polymeric human albumin) which recognize hepatitis B virus surface antigen. *Eur J Immunol* 1987; 17:365–370.
34. Colucci G, Beazer Y, Waksal SD: Interactions between hepatitis B virus and polymeric human serum albumin II. Development of syngeneic monoclonal anti–anti-idiotypes which mimic hepatitis B surface antigen in the induction of immune responsiveness. *Eur J Immunol* 1987; 17:371–374.

35. Uytdehaag FGCM, Osterhaus ADME: Induction of neutralizing antibody in mice against poliovirus type II with monoclonal antiidiotypic antibody. *J Immunol* 1985; 134:1225–1229.
36. McNamara MK, Ward RE, Kohler H: Monoclonal idiotype vaccine against *Streptococcus pneumoniae* infection. *Science* 1985; 226:1325–1326.
37. Grzych JM, Capron M, Lambert PH, Dissons C, Torres S, Capron A: An antiidiotypic vaccine against experimental schistosomiasis. *Nature* 1985; 316:74–76.
38. Weiss S, Bogen B: B-lymphoma cells process and present their endogenous immunoglobulin to major histocompatibility complex-restricted T cells. *Proc Natl Acad Sci USA* 1989; 86:282–286.
39. Waters SJ, Bona CA: Characterization of a T-cell clone recognizing idiotopes as tumor-associated antigens. *Cell Immunol* 1988; 111:87–93.
40. Rees ADM, Scoging A, Dobson N, Praputpittaya K, Young D, Ivanyi J, Lamb J: T cell activation by anti-idiotypic antibody: mechanism of interaction with antigen-reactive T cells. *Eur J Immunol* 1987; 17:197–201.

CHAPTER 3

Small Human V_H Gene Families Show Remarkably Little Polymorphism

Carol Williams, Linda Weigel, Inaki Sanz, and J. Donald Capra

After only a few human myeloma proteins were sequenced in the early 1960s, was it obvious that the human B cell repertoire displayed remarkable heterogeneity in that no two sequences were more than 85% identical. As murine proteins were sequenced in the late 1960s and early 1970s, more homogeneous structures were seen, and, as an understanding of the V, D, and J gene segments became apparent, it was obvious that, although rare, occasionally two independently derived murine myeloma protiens would be found with identical V_H regions. At the present time, with well over 300 human and murine myelomas and hybridomas sequenced, it is still very unusual to find two V_H regions that are identical.

These and other data have led to the conventional assumption that, especially in man (as an outbred species), immunoglobulin variable region gene segments differ considerably from individual to individual. That is, there is tremendous polymorphism in the human immune repertoire. These generalizations are supported by studies with restriction fragment length polymorphism analyses and, indeed, human gene sequences, which early suggested considerable variation particularly in germ line genes encoding $V_H I$, $V_H II$, and $V_H III$ gene families (1–3).

Within the last three years, three new human V_H families have been described; these are $V_H IV$, $V_H V$, and $V_H VI$. The $V_H IV$ family was first described by Lee et al as a V_H sequence adjacent to a $V_H III$ gene segment they had cloned using cosmid cloningvectors (4). Even though it cross-hybridized with $V_H II$ probes, the sequence was remarkably different from any previously described V_H gene or expressed protein to the point of defining a new V_H gene family. The prototype gene was isolated and used to probe genomic DNA and various phage and cosmid clones were isolated; the $V_H IV$ gene family was seen to be comprised of approximately eight V_H gene segments that were widely dispersed in the human V_H complex.

More recently, the $V_H V$ gene family was defined in a family with chronic lymphatic leukemia by Shen et al and Humphries et al (5, 6). This V_H gene family is quite small (two pseudogenes, one functional gene) and contains at least one member that is D_H proximal (6). At least one V_H gene is this family rearranges in a number of patients with acute and chronic lymphatic leukemia (6). Recently, a human anti-insulin antibody was found that differed in only a few aspects from $V_H 251$, the prototype $V_H V$ gene (7).

Almost simultaneously, groups in Seattle (8), England (9), and New York (10) isolated either cDNAs or genomic clones that clearly encoded yet a sixth gene family. This V_H gene "family" consists of but a single member. More remarkable was the fact that all three sequences isolated from three different individuals were identical, nucleotide for nucleotide (see below).

Our laboratory has long been interested in the structural basis of cross-reacting idiotypes and we set out to determine whether the smaller human V_H gene families were as polymorphic as the larger gene families appeared to be. A large body of experimental evidence in the inbred mouse suggests that whether large or small, gene families are quite polymorphic when one compares the V_H gene segments from one strain to another. Indeed, we are aware of only a few murine V_H genes isolated from one strain that are identical to those from another strain. In the human, we were quickly led to the conclusion that at least some human V_H genes must be remarkably preserved in evolution. Sanz et al (11) described an anti-Sm antibody that had an identical V_H sequence to two cDNA clones isolated from fetal liver by Schroeder et al (8) and a genomic clone isolated by Berman et al (10). The $V_H V$ anti-insulin antibody, also sequenced by Sanz et al (7), had a structure remarkably similar to $V_H V$ sequences (5, 6). Similarly, polyreactive antibodies demonstrated remarkable similarities among expressed $V_H III$ and $V_H IV$ genes to either cDNAs or to genomic structures previously described (7). We reasoned that a logical explanation for cross-reactive idiotypes in several human diseases and among several human antibodies might derive from the fact that the anti-idiotypic antibodies were directed toward framework structures on these small V_H families (12). As such, we wished to address the level of polymorphism among the human population in these families.

Methods

The polymerase chain reaction (PCR) reaction has allowed these studies to be done rather rapidly and easily. Since germ line gene structures were known for $V_H IV$ (4), $V_H V$ (5, 6), and $V_H VI$ (9, 10), it was relatively simple to construct priming oligonucleotides from the leader introns and heptamer–nonamer introns. Internal oligonucleotides were constructed to allow screening of both PCR reactions and, subsequently, cloned genes. A typical experiment involved 30 cycles of amplification, in which the priming oligonucleotides mentioned above were used: running the amplified mixture on an agarose gel, excising the band, and performing a fill-in reaction and then a blunt-end ligation into the *Eco*RV site of a phagemid vector. After appropriate growth in competent cells, lifts were probed with internal oligonucleotides, mini-preps were prepared, inserts were excised and examined, and those that were approximately 300 nucleotides in length were further characterized. Typically, single-stranded DNA was isolated from phagemids and both strands were sequenced by the dideoxy termination method (13) or by the 32-S ATP and a modified T7 DNA polymerase (Sequenase) or by a Dupont Genesis 2000 DNA sequencer with appropriate fluorescent didioxy nucleotides. All data were analyzed on a DNAStar computer system and sequences were compared using GeneBank.

Some $V_H IV$ Germ Line Genes Are Identical in Genetically Distinct Individuals

The PCR reactions were performed to study the germ line $V_H IV$ genes of two individuals. Approximately 15 clones were isolated from each individual and all 30 clones were sequenced. Most of these gene structures were similar to previously published sequences from Honjo's laboratory (4). Several of the sequences represented what are undoubtedly duplicate clones from the same individual. However, we found no structures from the same individual with sequences that differed by one or two nucleotides, which suggests that both alleles were identical. More remarkably, however, were the results shown in Figure 3.1, in which the identical nucleotide sequence was isolated from two different individuals. This sequence is 99.6% identical (only one nucleotide different) from the Honjo $V_H 11$ sequence. Therefore, three individuals exist in the world whose DNA for this particular $V_H IV$ gene is distinct by a minimum of a single nucleotide!

Some $V_H V$ Functional Genes as well as Pseudogenes Are Identical in Genetically Distinct Individuals

In the $V_H V$ family, we have isolated distinct functional $V_H V$ germ line genes from seven different individuals (Fig. 3.2). Two others are similar to the $V_H V$–$V_H 32$ described previously by Shen et al (5). These sequences are

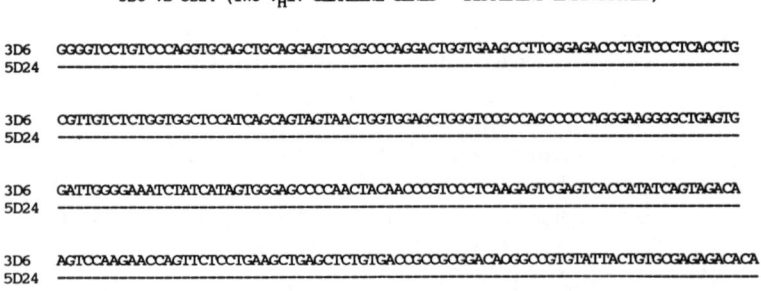

Fig. 3.1. These $V_H IV$ sequences from two genetically distinct individuals are identical and differ by only two nucleotides from a germ line $V_H IV$ ($V_H 11$) sequence previously published by Lee et al (4).

HUMAN VH5 SEQUENCES

```
VH251  GGAGTCTGTGCCGAGGTGCAGCTGGTGCAGTCTGGAGCAGAGGTGAAAAAGCCCGGGGAGTCTCTGAAGATCTCCTGTAAGGGTTCTGGATACAGCTTTA
VHJB   ----------------------------------------------------------------------------------------------------
VHCW   -----------------------------------G--------------------------------------------------------------
VHLB   ----------------------------------------------------------------------------------------------------
VHCH   ----------------------------------------------------------------------------------------------------
VHTT   ----------------------------------------------------------------------------------------------------
VHBLK  ----------------------------------------------------------------------------------------------------

VH251  CCAGCTACTGGATCGGCTGGGTGCGCCAGATGCCCGGGAAAGGCCTGGAGTGGATGGGGATCATCTATCCTGGTGACTCTGATACCAGATACAGCCCGTC
VHJB   ----------------------------------------------------------------------------------------------------
VHCW   ----------------------------------------------------------------------------------------------------
VHLB   ----------------------------------------------------------------------------------------------------
VHCH   ----------------------------------------------------------------------------------------------------
VHTT   ----------------------------------------------------------------------------------------------------
VHBLK  ----------------------------------------------------------------------------------------------------

VH251  CTTCCAAGGCCAGGTCACCATCTCAGCCGACAAGTCCATCAGCACCGCCTACCTGCAGTGGAGCAGCCTGAAGGCCTCGGACACCGCCATGTATTACTGT
VHJB   ----------------------------------------------------------------------------------------------------
VHCW   ---------------------------------------G------------------------------------------------------------
VHLB   ----------------------------------------------------------------------------------------------------
VHCH   ----------------------------------------------------------------------------------------------------
VHTT   ----------------------------------------------------------------------------------------------------
VHBLK  ----------------------------------------------------------------------------------------------------

VH251  GCGAGAGA
VHJB   --------
VHCW   --------
VHLB   --------
VHCH   --------
VHTT   --------
VHBLK  --------
```

Fig. 3.2. Six germ line genes isolated from six different individuals, including Caucasians, Blacks, and Orientals, are compared to $V_H 251$ (5,6). Their differences from the $V_H 251$ gene are indicated. Otherwise, identities are marked by dashes.

```
VH32   GGAGTCTGTCCGAAGTGCACCTGGTGCAGTCTGGACGAGAGGTGAAAAGCCCGGGGAGTCTCTGAGGATCTCTGTAAGGGTTCTGGATACAGCTTTAC
VHRG   ────X──────────────────────────────────────────────────C─────────────────────────────────────────
VHMW   ────X──────────────────────────────────────────────────C─────────────────────────────────────────

VH32   CAGCTACTGGATGGCTGGGTGCGCCAGATGCCCGGGAAAGGCCTGGAGTGGATGGGGGAGATTGATCCTAGTGACTCTTATACCAACTACAGCCGGTCC
VHRG   ───────────────────────────A─────────────────────────────────────────────────────────────────────
VHMW   ───────────────────────────A─────────────────────────────────────────────────────────────────────

VH32   TTCCAAGGCCAGGTCACCATCTCAGCTGACAAGTCCATCAGCACTGCCTACCTGCAGTGGAGCAGCCTGAAGGCCTCGGACACCGCCATGTATTACTGTG
VHRG   ─────────────────────────────────────────────────────────────────────────────────────────────────
VHMW   ─────────────────────────────────────────────────────────────────────────────────────────────────

VH32   CCAGA
VHRG   ─────
VHMW   ─────
```

Fig. 3.3. Comparison of PCR-amplified sequences from two genetically unrelated individuals to the previously described V_H32 gene (5, 6).

displayed in Figures 3.2 and 3.3. Figure 3.2 shows that among seven different individuals, there are only three nucleotide differences in the 300 base pairs that can be compared. This represents a level of variation (3 divided by 7 times 306) of 0.1%. This same degree of preservation is evident in the $V_H 32$ genes illustrated in Figure 3.3. The three $V_H 32$ genes sequenced from three different individuals differ by a maximum of two nucleotides, which indicates a level of variation of 0.2% in this sample. This remarkable degree of homogeneity among 10 different germ line genes of the $V_H V$ family was wholly unexpected.

Three $V_H VI$ Germ Line Genes Isolated in Three Different Laboratories Are Identical and Also Are Identical to Three Additional $V_H VI$ Gene Segments Sequenced in Our Laboratory

As mentioned above, Alt, Perlmutter, and Rabbitts independently sequenced human $V_H VI$ genes. Within the coding segments, all three of these genes are identical. Using PCR reactions we have isolated three $V_H VI$ genes from different individuals and found that they also are identical (Fig. 3.4).

Still to be determined is whether the $V_H I$, II, and III gene families show the same level of homogeneity. Indeed, the human anti-Sm antibody sequenced by Sanz et al (11) as well as two anti-DNA antibodies sequenced by Dersimonian et al (14) suggest that at least in some $V_H III$ members there will also be this preservation of structure.

Discussion

How can we reconcile these data with previously held notions? First, most myeloma proteins and, indeed, most cDNA sequences that have been done are on IgG antibodies. These likely reflect some level of somatic mutation and as such we may all have been misled by these variations. Second, myeloma proteins themselves may be subject to an increased level of somatic mutation, since there is growing evidence that the system for somatic mutation does not turn off even in the fully differentiated plasma cells that are characteristic of myelomas. Third, the phenomenon might be one that is limited to the smaller human V_H families and not to be larger V_H families. We have called some of the $V_H III$ sequences mentioned above $V_H IIIb$ because their sequence is actually on the borderline between the same and a different V_H family (11). For example, 4B4 (the prototype $V_H IIIb$ sequence) is identical to two cDNA sequences of Schroeder et al (8) and, at the same time, no other $V_H III$ sequence is similar to it by more than 80%. Typically, V_H families are defined as more than 80% different but, more often than not, they are more than 70% different. Thus, these structures at the border line may represent a small gene family.

HUMAN V$_H$VI SEQUENCES

```
                  GGTGTCTCTCACAGGTACAGCTGCAGCAGTCAGGTCCAGGACTGGTGAAGCCCTCGCAGACCCTCACTCAACTGCATCCTCCGGGGACAGTGTCTCT
V$_H$VI (ref 9)
V$_H$VI (ref 8)   ----------T-------------------------------------------------------------------------------------
V$_H$VI (ref 11)  ----------T-------------------------------------------------------------------------------------

                  AGCAACAGTGCTGCTTTGAACTGGATCAGGCAGTCCCCATCGAGAGGCCTTGAGTGGCTGGAAGGACATACAGTCCAAGTGTATAATGATTATGCAGT
V$_H$VI (ref 9)
V$_H$VI (ref 8)
V$_H$VI (ref 11)

                  ATCTGTGAAAAGTCGAATAACCATCAACCCAGACACATCCAAGAACCAGTTCTCCCTGCAGCTGAACTCTGTGACTCCCGAGGACACGGCTGTGTATTACTGT
V$_H$VI (ref 9)
V$_H$VI (ref 8)
V$_H$VI (ref 11)

                  GCAAGA
V$_H$VI (ref 9)
V$_H$VI (ref 8)   ------
V$_H$VI (ref 11)  ------
```

Fig. 3.4. Three V$_H$VI genes isolated in different laboratories (8, 9, 11) are identical.

Viewed in this light, the human V_H genes appear no different than the human genes for any gene family be they pancreatic elastases, albumin, globin, or the low density lipoprotein (LDL) receptor. In these instances, exonic sequences are known to vary at most at the level of 0.1%, which is approximately the level we are seeing here. The notion that human V_H genes are quite polymorphic is, therefore, incorrect and it may well be that all human antibody genes are as similar to each other as are all other genes in the human germ line. Thus, the remarkable variation that is seen in the human population among expressed antibodies would have to be explained by somatic mutation. It may well be that the reason this level of germ line variation has not been seen in the murine system is because most murine antibodies that have been studied are directed toward small haptens and the techniques used for the isolation of hybridomas differ from the human work. Nonetheless, the findings reported above suggest a remarkable homogeneity of human V_H sequences (particularly among the smaller human V_H families), and these similar structures may explain the structural basis of cross-reacting idiotypes among human antibodies, particularly autoantibodies that typically are IgM and much closer to germ line.

Acknowledgments. The authors acknowledge the skilled technical assistance of Stephen Scholl and Maggy Fina. We are grateful to Dr. Kathy Meek for many helpful discussions and to Dr. Philip Tucker for not only providing the prototype $V_H V$ germline sequence but also for many stimulating and critical comments. This work was supported in part by the National Institutes of Health (AR 39169) and a Grant from the Tobacco Research Institute. We thank Margaret Wright for skilled secretarial assistance.

References

1. Matthyssens G, Rabbitts TH: Structure and multiplicity of genes for the human immunoglobulin heavy chain variable region. *Proc Natl Acad Sci USA* 1980; 77:6561.
2. Honjo T: Immunoglobulin genes. *Ann Rev Immunol* 1983; 1:499.
3. Rechavi G, Ram D, Glazer L, Zakut R, Givol D: Evolutionary aspects of immunoglobulin heavy chain variable region (V_H) gene subgroups. *Proc Natl Acad Sci USA* 1983; 80:855.
4. Lee KH, Matsuda F, Knashi T, Kodaira M, Honjo T: A novel family of variable region genes of the human immunoglobulin heavy chain. *J Molec Biol* 1987; 195:1.
5. Shen A, Humphries C, Tucker P, Blattner F: Human heavy-chain variable region gene family nonrandomly rearranged in familial chronic lymphocytic leukemia. *Proc Natl Acad Sci USA* 1987; 84:1987.
6. Humphries CG, Shen A, Kuziel WA, Capra JD, Blattner FR, Tucker PW: Characterization of a new human immunoglobulin V_H family that shows preferential rearrangement in immature B cell tumors. *Nature (Lond)* 1988; 331:446.

7. Sanz I, Casali P, Thomas JW, Notkins AL, Capra JD: Genetics basis of natural autoantibodies: organization, complexity, and mechanisms of diversity of the human B cell repertoire. *J Immunol* 1989; 142:4054.
8. Schroeder HW, Hillson JL, Perlmutter RM: Early restriction of the human antibody repertoire. *Science* 1987; 238:791.
9. Buluwela L, Rabbitts TH: A V_H gene is located within 95 Kb of the human immunoglobulin heavy chain constant region genes. *Eur J Immunol* 1988; 18:1843–1845.
10. Berman JE, Mellis SJ, Pollock R, et al: Content and organization of the human Ig V_H locus: Definition of three new V_H families and linkage to the Ig C_H locus. *EMBO J* 1988; 7:727–738.
11. Sanz I, Dang H, Takei M, Talal N, Capra JD: V_H sequence of a human anti-Sm autoantibody: Evidence that autoantibodies can be unmutated copies of germline genes. *J Immunol* 1980; 142:883.
12. Makar R, Sanz I, Thomas JW, Capra JD: A structural basis for human cross-reacting idiotypes. *Ann Inst Pasteur/Immunol* 1988; 139:651.
13. Sanger F, Nicklen S, Coulson AR: DNA sequencing with chain terminating inhibitors. *Proc Natl Acad Sci USA* 1977; 74:5463.
14. Dersimonian H, Schwartz RS, Barrett KJ, Stollar BD: Relationship of human variable region heavy chain germline genes to genes encoding anti-DNA autoantibodies. *J Immunol* 1987; 139:2496.

CHAPTER 4

Anti-Idiotypic Vaccines for Carbohydrate Antigens

Kathryn E. Stein

Introduction

In this chapter I will review the literature in which anti-idiotypic (anti-Id) antibodies have been used to induce or prime for an antibody response to a carbohydrate determinant, whether a polysaccharide or an oligosaccharide side chain of a glycoprotein or a synthetic conjugate. In each case the subject will be discussed in terms of the ability of anti-idiotypes to stimulate a response and in terms of the data presented that suggest that the anti-idiotype might behave as an internal image of antigen or Ab2β (1). The criteria I will use to determine whether an anti-idiotype is an internal image are those first discussed with reference to vaccines by Nisonoff and colleagues (2, 3).

These include the ability of anti-idiotypes to react with antibodies of a given specificity from different inbred strains and different species, to react with most of a population of those antibodies, and to stimulate antibodies of the given fine specificity. The question of whether a globular protein can resemble a polysaccharide is one that has intrigued many investigators. Although there are no structural data that answer the question at this time, there are some data that suggest that anti-idiotypes can function as internal image or antigen mimic vaccines for carbohydrates. In some instances, data show that anti-idiotypes can induce an immune response to a carbohydrate determinant in situations in which antigen, itself, is unable to induce the response. From the perspective of vaccine development, an Ab2β need not be a requirement for a successful vaccine, although the genetic constraints (4-7) of some conventional anti-Id or AB2α vaccines (1) may preclude their use in an outbred population such as man. There may be much more conservation of immunoglobulin (Ig) Variable (V) regions in man, however, than might have been anticipated from studies in inbred animals. Cross-reactive idiotypes (CRI) recently have been reported from several laboratories on human anti-*Haemophilus influenzae* type b polysaccharide antibodies from unrelated individuals (8-10). Such shared idiotypes have been reported previously for human anti-tetanus antibodies (11, 12), anti-streptococcal group A polysaccharide antibodies (13), and antibodies to the mannan polysaccharide of *Candida albicans* (14). Thus, anti-idiotypes prepared against antibodies from a small number of unrelated individuals, even if not of the antigen mimic type, might prove to be highly effective and should not be ruled out as vaccine candidates.

What are the potential advantages of anti-Id vaccines over conventional vaccines for the development of anti-carbohydrate antibody responses? The most obvious and most important is that they might work when antigen, itself, does not. This is particularly relevant for immunization of the neonate, as the ability to respond to polysaccharide antigens develops late in ontogeny (15–19). Conversion of these thymus-independent (TI) anti-polysaccharide (PS) responses to thymus-dependent (TD) responses by the use of poly- or oligosaccharide–protein conjugate vaccines can shift the age-related response curve to much younger ages (20–25). Another way to achieve a TD response might be to use an anti-idiotype. In at least one system, anti-Id vaccination has been shown to require the presence of the thymus (26). Data to be reviewed below on the meningococcal group C anti-Id vaccine (27) indicate that anti-idiotypes stimulated a memory response, whereas PS, as expected, did not; this suggests that here, too, anti-idiotype behaved as a TD antigen. Anti-idiotype might also be used as a TD antigen to stimulate an anti-PS antibody response in individuals with selective immunodeficiencies who are unresponsive to PS, itself, but who can respond to TD antigens after repeated immunizations (28).

Another reason to use anti-idiotypes might be the lack of purified, well-characterized conventional antigen. Although recombinant DNA technology may eliminate some of these problems for protein antigens, the problems for polysaccharides or oligosaccharide chains of glycoproteins, which are not the direct products of the structural genes, will be less readily solved. It is possible to generate monoclonal antibodies (idiotypes) that can be shown to confer passive protection in an appropriate model system, yet the actual protective antigen may not have been isolated in pure form. Anti-Id vaccines could be generated that would induce protective immunity and could substitute for antigen. Finally, anti-idiotypes could be generated against specific idiotypes, which are known to be highly protective. Thus, the selectivity of anti-idiotypes could be advantageous in stimulating protective immunity, the ultimate goal of any vaccine.

A limited number of published experiments involved the use of anti-idiotypes to stimulate antibodies directed against carbohydrate epitopes. These will be discussed in chronological order in the sections below. Throughout this chapter, I will use the terms Ab1 to refer to antibody directed at a nominal antigen, Ab2α and Ab2β to refer to anti-idiotypes of the conventional and internal image types, respectively, and Ab1' to refer to the antigen-binding antibodies induced by Ab2 (1).

Bacterial Levan

Bacterial levan (BL) is a branched chain of polyfructosan consisting of a $\beta(2 \to 6)$ linked fructosan backbone with $\beta(2 \to 1)$ linkages at the branch points. Lieberman et al (29) described a cross-reactive idiotype expressed on a large portion of BALB/c anti-BL antibodies and some myeloma proteins, including E109 and UPC-61, which have specificity for $\beta(2 \to 1)$ linked fructose. Another idiotype, Id A48, is expressed on the A48 myeloma protein, which is specific for $\beta(2 \to 6)$ linked fructose and is a so-called silent idiotype in the BALB/c response to BL in that only 5 to 10% of BL plaque-forming cells (PFC) express it. Hiernaux et al (30) demonstrated that low doses of anti-A48, when given to 1-day-old mice, resulted in the expression of Id A48 on 45 to 73% of anti-BL PFC. Thus, administration of anti-idiotype at birth resulted in the subsequent activation of a normally silent clone upon immunization with BL.

These studies were extended by Rubinstein et al (31). They examined the ability of several monoclonal anti-Id A48 antibodies (mAbs) to substitute for antigen in the induction of idiotype A48 following neonatal priming with polyclonal anti-A48. They found that one of six mAbs tested, 17-38, could substitute for the antigen requirement in the activation of anti-BL PFC. The response to 17-38 was dose dependent and resembled the unprimed response to BL in adult mice in that A48 was a minor component of the PFC response and the serum antibodies had spectrotypes, by isoelectric focusing analysis, that were similar to those of the antibodies induced by BL. In addition, BL was shown to inhibit the binding of 17-38 to A48. By these criteria, 17-38 appears to function as an antigen mimic or Ab2β. The ability of 17-38 to induce anti-BL in species other than mice has not been tested, however. The explanation for the differential effects of anti-A48 priming on the subsequent response to 17-38 and BL remains unclear.

Another aspect of neonatal anti-idiotype treatment examined by Rubinstein et al (31) was which cells were actually primed. They showed that purified B cells taken from 1-month-old mice that had been given anti-A48 at birth were sufficient to transfer an id A48-dominant response to naive recipients. These experiments did not, however, address the role of T cells in the initial priming. Other studies have suggested that anti-idiotype induction is a TD process. One study showed that nude mice do not make an anti-RNAse response after anti-idiotype treatment (26), another study that adoptive transfer of T cells was associated with an increase in idiotypes (32), and a third (27), which will be discussed below, that anti-idiotype induction of anti-meningococcal group C antibodies had the features of a memory response.

Escherichia coli K13

Following observations made by Hiernaux et al (30) and by Rubinstein et al (31), that anti-idiotype could be used to prime the immune system for a subsequent response to a PS antigen, Tommy Söderström and I undertook an investigation to determine whether such a primed response for a capsular PS of a bacterial pathogen, *E. coli* K13, could result in protective immunity upon subsequent challenge with a lethal dose of bacteria (33). The PS capsule of *E. coli* K13 has a relatively simple structure, consisting of a disaccharide repeat unit of 3)-β-ribofuranosyl-$(1 \to 7)$-β-KDO-2($\to$ with O-acetyl groups on carbon 4 of the KDO (34). A series of BALB/c mAbs were produced against the K13 PS (35). One mAb, 150C8, which was demonstrated to be protective in a passive protection system (36), was used to produce an allogeneic IgG1 anti-idiotype, 5858C, in A/He mice. The id 150C8 detected by 5868C was not detected on any other member of a small panel of anti-K13 MAbs. However, id 150C8 was shown to be present in sera from different strains of mice (BALB/c, CBA, and C57BL/6) and was shown to increase in both BALB/c and CBA mouse sera following immunization with K13. In addition, id 150C8 was detected in the serum of two of three rats that had been hyperimmunized with K13 vaccine but not in normal rat serum.

The model used to test 5868C as a vaccine involved priming with anti-idiotype within 24 hours after birth, immunizing at 4 weeks of age with a whole cell K13 vaccine or purified K13 PS, and challenging 1 week later with 20 to 30 LD_{50} K13 bacteria. Mice immunized at 4 weeks of age without neonatal priming, as well as unimmunized mice, survived the challenge poorly (0 to 44% survival), whereas mice treated at birth with 50 ng of 5868C were protected from lethal challenge (78 to 93% survival). Furthermore, the effects of anti-idiotype priming were long lived. Thus when immunization was delayed until the mice were 12 weeks of age, and then followed 1 week later by a challenge of 50 LD_{50}, 78% of mice survived compared to 0 to 25% of the control animals. In this experiment 50 ng but not 1 μg of 5868C primed for a subsequent protective response. This is similar to the dose response reported by Kelsoe et al (37) for anti-idiotype induction of the B1-8 idiotype found on anti-NP antibodies of C57BL/6 mice.

Although no direct proof was obtained that 5868C is an Ab2β, several results provide suggestive evidence that it is. These include the ability of K13 PS to inhibit the 5868C–150C8 interaction, reactivity of 5868C with antibodies from different mouse strains and different species (rat), the increase in reactivity of 5868C with sera from mice immunized with K13, and the ability of 5868C to prime neonatal cells. The last feature is presumed to be due to the TD nature of anti-idiotype priming, as the TI antigen K13 PS did not prime for a protective response.

Trypanosoma cruzi Surface Glycoprotein

GP72 is a 72kD glycoprotein containing 49% carbohydrate that is expressed on the surface of *T. cruzi*. This molecule as well as a mAb, W1C 29.26, which is specific for the carbohydrate portion of GP72, have been characterized by Snary and colleagues (38, 39). Sacks et al (40) used WIC 29.26 to generate Ab2 in rabbits to be used as a vaccine for *T. cruzi*. BALB/c mice were immunized with affinity-purified rabbit Ab2 and the serum from such mice was shown to bind to the Tulakuen strain of *T. cruzi* but not to the Brazil strain, which expresses a GP72 that lacks the surface-expressed carbohydrate. Thus the induced antibodies, just as was WIC 29.26, were specific for the carbohydrate epitopes. The BALB/c mice immunized with Ab2 and boosted with *T. cruzi* showed a greatly increased expression of id WIC 29.26. Also, Ab2 induced anti-GP72 in rabbits and guinea pigs. It reacted with affinity-purified rabbit anti-GP72 and with human sera from *T. cruzi-* but not *Leishmania donovani-*infected individuals. By these criteria the affinity-purified rabbit Ab2 is either an Ab2β bearing an internal image of the carbohydrate epitopes of GP72 or it detects a shared idiotype present on the anti-GP72 antibodies of mice, rabbits, guinea pigs, and man. Lacking sequence data of anti-GP72 antibodies from these different species, the possibility of an interspecies idiotype cannot be ruled out. As mentioned above and as was pointed out by Sacks et al (40), whether an anti-Id vaccine bears an internal image of antigen or not, the important consideration is whether the Ab2 functions as antigen. In this regard, WIC 29.26 Ab2 mimics the properties of GP72. The induction of anti-GP72 with Ab2 was not sufficient to protect mice from *T. cruzi* infection. As Sacks et al point out, however, WIC 29.26 itself was not able to confer passive protection.

Streptococcal Group A Carbohydrate

The immunodominant specificity of *Streptococcus pneumoniae* group A carbohydrate (A-CHO) is *N*-acetyl-D-glucosamine (GlcNAc), which is linked to a rhamnose backbone (41). The antibody response to A-CHO, like that to other PS, is clonally restricted in both mouse and man (13, 42). Bloem et al (43) have raised a monoclonal anti-idiotype against id 498. Anti-id 498 detects a recurrent idiotype expressed on human IgM anti-GlcNAc antibodies. The id 498 was detected in 19 of 27 sera from randomly chosen individuals. Anti-id 498 coupled to agarose beads was able to induce IgM anti-A-CHO in cultures of purified B cells plus T cell-derived soluble factors from four of six normal donors who expressed serum id 498. This example of anti-idiotype induction of anti-PS antibodies differs from the other examples cited here in that anti-id 498 appears to be of the Ab2α type, which detected id 498 in 70% of individuals tested and presumably functions by direct stimulation of idiotype on B cells. If other recurrent idiotypes are detected in such high frequency, however, then even a small mixture of Ab2α might provide an adequate vaccine. As discussed in the introduction, shared idiotypes among unrelated individuals have been reported on human antibodies to a number of foreign antigens.

Neisseria meningitidis Group C Polysaccharide

The group C PS (MCPS) of *N. meningitidis* is a homopolymer of $\alpha(2 \rightarrow 9)$ linked sialic acid. Westerink et al (27) used a monoclonal IgM anti-MCPS, 1E4, as the selecting antigen for anti-idiotypes. Their approach to anti-idiotype production was somewhat unusual in that the splenocytes used in the fusion to produce a BALB/c syngeneic monoclonal anti-idiotype, 6F9, were derived from a mouse carrying the ascites tumor of 1E4. Several pieces of data suggest that 6F9 may be an Ab2β. Of six anti-idiotypes tested, 6F9 was the only one that inhibited the binding of 1E4 to MCPS, suggesting that the idiotype recognized by 6F9 is in or near the combining site. Rabbits immunized with 6F9-keyhole limpet hemocyanin (a TD carrier protein) and boosted with 6F9 had significant titers of anti-MCPS antibodies. This boosted anti-MCPS response was taken as evidence of the TD nature of the anti-idiotype stimulation, as two doses of TI MCPS did not boost the response. In addition, cells producing 6F9 antibodies were shown to bind rabbit anti-MCPS typing antiserum but not control, rabbit anti-group A, typing serum. Furthermore, 6F9 inhibited the binding of human anti-MCPS antibodies from a patient convalescing from meningococcal group C sepsis to MCPS is an ELISA assay. The investigators also claimed that 6F9 could induce Ab1' in BALB/c mice with specificity similar to Ab1. Based on studies by Rubinstein and Stein (44), however, in which both IgM and IgG anti-MCPS antibodies in pre-immune sera from virtually all BALB/c mice tested were demonstrated, the presence of anti-MCPS in sera from mice not deliberately immunized with MCPS or meningococcal group C vaccine, alone, cannot be taken a proof of the inducing effects of 6F9. Nonetheless, the aggregate of data presented indicates that 6F9 can be used as an anti-idiotype vaccine.

3-O-α-L-Fucopyranosyl-β-D-Galactopyranoside (Fucα(1 → 3)Gal)

Fucα(1 → 3)Gal is a carbohydrate epitope that has been shown to be a tumor-associated antigen. Diakun and Matta (45) have used this determinant as the basis for an anti-idiotype vaccine. The Ab1 consisted of affinity-purified polyclonal rabbit antibodies raised against a Fucα(1 → 3)Gal-BSA conjugate. Rabbit Ab2 antibodies were affinity-purified on an Ab1 column. By ELISA inhibition assays, the investigators showed that Ab2 could inhibit the binding of Ab1 to antigen and that the Ab1–Ab2 interaction could be inhibited by antigen. Purified Ab2 was used to immunize a third rabbit and the Ab1' produced by this rabbit was purified on an antigen affinity column. Because of the well-defined specificity of the Ab1, the investigators were able to use a series of closely related antigens to examine and compare the specificities of Ab1 and Ab1'. They showed that both antibody populations had similar binding curves on antigen-coated plates and similar inhibition profiles by four different antigens that contained Fucα(1 → 3)Gal. Neither population of antibodies was inhibited by antigens such as Fucβ(1 → 3)Gal or Fucα(1 → 2)Gal. These data indicate that Ab2 in this system can function like antigen, but they do not address the issue of reactivity of Ab2 with antibodies from different species.

Conclusions

In this chapter, I have reviewed the studies in which an Ab2 has been raised against an Ab1 with specificity for a carbohydrate determinant. Although small in number, these studies cover a variety of systems that involve poly- and oligosaccharide antigens and mono-and polyclonal Ab1 and Ab2. In all of these systems, Ab2 was able to induce an antigen-specific response. The mechanisms involved in each system, however, are likely to be as different as the systems themselves.

Very little attention has been given by any of the investigators to the mechanism of anti-idiotype induction or the salient features of a given system that would be useful in another. It is fair to say that the rules for generating "successful" Ab2 have not been developed, nor is it clear that it will be possible to do so. Recently Misplon et al (46) examined a series of monoclonal Ab2 for such immunochemical parameters as fine specificity, relative affinity, and isotype, with the aim of attempting to predict from these properties which Ab2s would be good inducers of Ab1'. They found dramatic differences in the abilities of the Ab2s to induce Ab1', but none of the measured parameters correlated with differences in induction.

Another question of particular importance for the development of Ab2 vaccines is whether an antibody against a nominal antigen (Ab1) must be protective in order for an Ab2 to stimulate protective immunity. I believe this subject has not been systematically investigated. It would seem prudent to select an Ab1 that has been demonstrated to confer passive protection (assuming that a suitable model for protection exists) when selecting antibodies for the production of Ab2 vaccines, but this warrants further investigation.

In summary, a number of different carbohydrate antigen systems have been utilized for the production of Ab2 that can induce anti-carbohydrate Ab1'. Some antibodies appear to be Ab2β, whereas others are likely to be Ab2α, but both types can function as inducing antibodies. More systematic investigation in several systems will be required before principles are established that allow one to predict the requirements for an inducing Ab2. Direct structural evidence of an antigen mimic for a poly- or oligosaccharide will not be possible, unlike the reovirus system (47, 48) in which both the antigen and the anti-idiotype are proteins and can be sequenced. In the case of carbohydrate systems, proof will require a determination of the three-dimensional structures of both the Ab2 and the carbohydrate antigen. The limitations for development of clinically useful vaccines, however, lie not so much in the proof of Ab2β structure or even the determination of the rules for generating functional Ab2 but rather in the determination of the dominant and relevant idiotypes on human antibodies specific for a given polysaccharide antigen. It is hoped that progress in this area will be rapid.

Acknowledgments. I am grateful to Ms. Julia Misplon and Dr. Suzanne Epstein for their critical review of the manuscript.

References

1. Jerne NK, Roland J, Cazenave P-A: Recurrent idiotopes and internal images. *EMBO J* 1982; 1:243–247.
2. Nisonoff A, Lamoyi E: Implications of the presence of an internal image of the antigen in anti-idiotypic antibodies: Possible application to vaccine production. *Clin Immunol Immunopathol* 1981; 21:397–406.
3. Gurish MF, Nisonoff N: Potential use of anti-idiotype antibodies as vaccines, in Dreesman GR, Bronson JG, Kennedy RC, eds: *Proc First Ann Southwest Foundation for Biomedical Research Int Symp Houston Texas Nov 8–10 1984.* (American Society of Microbiology, Washington, D.C., 1985) p. 103.
4. Takemori T, Tesch H, Reth M, Rajewsky K: The immune response against anti-idiotope antibodies. I. Induction of idiotope-bearing antibodies and analysis of the idiotope repertoire. *Eur J Immunol* 1982; 12:1040–1046.
5. Epstein SL, Masakowski VR, Sharrow SO, Bluestone JA, Ozato K, Sachs DH: Idiotypes of anti-Ia antibodies. II. Effects of in vivo treatment with xenogeneic anti-idiotype. *J Immunol* 1982; 129:1545–1552.
6. Trenkner E, Riblet R: Induction of antiphosphorylcholine antibody formation by anti-idiotypic antibodies. *J Exp Med* 1975; 142:1121–1132.
7. Sacks DL, Sher A: Evidence that anti-idiotype induced immunity to experimental African trypanosomiasis is genetically restricted and requires recognition of combining site-related idiotypes. *J Immunol* 1983; 131:1511.
8. Lucas AH: Expression of crossreactive idiotypes by human antibodies specific for the capsular polysaccharide of *Hemophilus influenzae* B. *J Clin Invest* 1988; 81:480–486.
9. Lucas AH, Granoff DM: A major cross-reactive idiotype associated with human antibodies to the capsular polysaccharide of *H. influenzae* B; expression in relation to age and vaccine. *FASEB J* 1989; 3:A1270-A1270. Abstract.
10. Rosen SM, Allen C, Waksal HW: Presence of internal image on murine monoclonal anti-idiotypes to human antibody directed against polyribosyl-ribitol phosphate (PRP) of *Haemophilus influenzae* type B. *FASEB J* 1989; 3:A674-A674. Abstract.
11. Altevogt P, Wigzell H: A V_H-Associated idiotype in human anti-tetanus antibodies. *Scand J Immunol* 1983; 17:183–192.
12. Hoffman WL, Strucely PD, Jump AA, Smiley JD: A restricted human antitetanus clonotype shares idiotypic cross-reactivity with tetanus antibodies from most human donors and rabbits: Reactivity with antibodies of widely differing electrophoretic mobility. *J Immunol* 1985; 135:3802–3807.
13. Emmrich F, Zenke G, Eichmann K: Isotype restriction of idiotopes associated with human anti-streptococcal A carbohydrate antibodies. *Eur J Immunol* 1986; 16:542–546.
14. de Saint Basile G, Durandy A, Somme G, Griscelli C: Idiotypy of human anti-*Candida albicans* antibodies: Recurrence, presence of a cross-reactive autoanti-idiotypic-like activity, and role in the induction of specific in vitro antibody response. *J Immunol* 1987; 138:417–422.
15. Kayhty H, Karanko V, Peltola H, Makela PH: Serum antibodies after vaccination with *Haemophilus influenzae* type B capsular polysaccharide and responses to reimmunization: No evidence of immunologic tolerance or memory. *Pediatrics* 1984; 74:857–865.

16. Parke JC Jr, Schneerson R, Robbins JB, Schlesselman JJ: Interim report of a controlled field trial of immunization with capsular polysaccharides of *Haemophilus influenzae* type b and group C *Neisseria meningitidis* in Mecklenburg County, North Carolina (March 1974–March 1976). *J Infect Dis* 1977; 136S:S51–S56.
17. Peltola H, Kayhty H, Sivonen A, Makela PH: *Haemophilus influenzae* type capsular polysaccharide vaccine in children: A double-blind field study of 100,000 vaccinees 3 months to 5 years of age in Finland. *Pediatrics* 1977; 60:730–737.
18. Smith DH, Peter G, Ingram DL, Harding AL, Anderson P: Responses of children immunized with the capsular polysaccharide of *Hemophilus influenzae*, type b. *Pediatrics* 1973; 52:637–644.
19. Bona C, Mond JJ, Stein KE, House S, Lieberman R, Paul WE: Immune response to levan. III. The capacity to produce anti-inulin antibodies and cross-reactive idiotypes appears late in ontogeny. *J Immunol* 1979; 123:1484–1490.
20. Stein KE, Zopf DA, Miller CB, et al: The immune response to a thymus-dependent form of B512 Dextran requires the presence of Lyb5$^+$ lymphocytes. *J Exp Med* 1983; 157:657–666.
21. Schneerson R, Robbins JB, Chu C, et al: Serum antibody responses of juvenile and infant rhesus monkeys injected with *Haemophilus influenzae* type B and pneumococcus type 6A capsular polysaccharide-protein conjugates. *Infect Immun* 1984; 45:582–591.
22. Eskola J, Kayhty H, Peltola H, et al: Antibody levels achieved in infants by course of *Haemophilus influenzae* type b polysaccharide/diphtheria toxoid conjugate vaccine. *Lancet* 1985; 1:1184–1186.
23. Anderson P, Pichichero ME, Insel RA: Immunization of 2-month-old infants with protein-coupled oligosaccharides derived from the capsule of *Haemophilus influenzae* type b. *J Pediatr* 1985; 107:346–351
24. Lepow ML, Samuelson JS, Gordon LK: Safety and immunogenicity of *Haemophilus influenzae* type b-polysaccharide diphtheria toxoid conjugate vaccine in infants 9 to 15 months of age. *J Pediatr* 1985; 106:185–189.
25. Insel RA, Anderson PW: Oligosaccharide-protein conjugate vaccines induce and prime for oligoclonal IgG antibody responses to the *Haemophilus influenzae* b capsular polysaccharide in human infants. *J Exp Med* 1986; 163:262–269.
26. Miller GG, Nadler PI, Asano Y, Hodes RJ, Sachs DH: Induction of idiotype-bearing nuclease-specific helper T cells by in vivo treatment with anti-idiotype. *J Exp Med* 1981; 154:24–34.
27. Westerink MAJ, Campagnari AA, Wirth MA, Apicella MA: Development and characterization of an anti-idiotype antibody to the capsular polysaccharide of *Neisseria meningitidis* serogroup C. *Infect Immun* 1988; 56:1120–1127.
28. Insel RA, Anderson PW: Response to oligosaccharide–protein conjugate vaccine against *Hemophilus influenzae* b in two patients with IgG2 deficiency unresponsive to capsular polysaccharide vaccine. *N Engl J Med* 1986; 315:499–503.
29. Lieberman R, Potter M, Humphrey W Jr, Mushinski EB, Vrana M: Multiple individual and cross-specific idiotypes on 13 levan-binding myeloma proteins of BALB/c mice. *J Exp Med* 1975; 142:106–119.
30. Hiernaux J, Bona C, Baker PJ: Neonatal treatment with low doses of anti-idiotype antibody leads to the expression of a silent clone. *J Exp Med* 1981; 153:1004–1008.
31. Rubinstein LJ, Goldberg B, Hiernaux J, Stein KE, Bona CA: Idiotype-antiidiotype Regulation V. The requirement for immunization with antigen or monoclonal antiidiotype antibodies for the activation of beta(2-6) and beta(2-1) polyfructosan-

reactive clones in BALB/c mice treated at birth with minute amounts of anti-A48 idiotype antibodies. *J Exp Med* 1983; 158:1129–1144.
32. Bluestone JA, Auchincloss H Jr, Epstein SL, Sachs DH: Idiotypes of anti-MHC monoclonal antibodies. In: Kohler H, Urbain J, Cazenave P-A eds: *Idiotypy in Biology and Medicine*. Orlando, Fla: Academic Press; 1984: p 243.
33. Stein KE, Söderström T: Neonatal administration of idiotype or antiidiotype primes for protection against *E. coli* K13 infection in mice. *J Exp Med* 1984; 160:1001–1011.
34. Vann WF, Söderström T, Egan W, et al: Serological, chemical, and structural analyses of the *Escherichia coli* cross-reactive capsular polysaccharides K13, K20, and K23, *Infect Immun* 1983; 39:623–629.
35. Söderström, T, Stein K, Brinton CC Jr, et al: Analysis of *Escherichia coli* K1, K13 and Type 1 pilus antigens with monoclonal antibodies. In: Hanson LÅ, ed: *Progress in Allergy* Vol. 33. Basel: Karger; 1983:259.
36. Söderström T, Brinton CC Jr, Fusco P, et al: Analysis of pilus mediated pathogenic mechanisms with monoclonal antibodies. In: Schlessinger D ed: *Microbiology 1982* Washington, D.C., ASM Publications; 1982: p 305.
37. Kelsoe G, Reth M, Rajewsky K: Control of idiotope expression by monoclonal anti-idiotope and idiotope-bearing antibody. *Eur J Immunol* 1981; 11:418–423.
38. Snary D, Ferguson MAJ, Scott MT, Allen AK: Cell surface antigens of *Trypanosoma cruzi*: Use of monoclonal antibodies to identify and isolate a epimastigote specific glycoprotein. *Mol Biochem Parasit* 1981; 3:343.
39. Ferguson MAJ, Allen AK, Snary D: Studies on the structure of phosphoglycoprotein from the parasitic protozoan *Trapanosoma cruzi*. *J Biochem* 1983; 213:313.
40. Sacks DL, Kirchhoff LV, Hieny S, Sher A: Molecular mimicry of a carbohydrate epitope on a major surface glycoprotein of *Trypanosoma cruzi* by using anti-idiotypic antibodies. *J Immunol* 1985; 135:4155–4159.
41. Coligan JE, Schnute WC Jr, Kindt TJ: Immunochemical and chemical studies on streptococcal group-specific carbohydrates. *J Immunol* 1975; 114:1654–1658.
42. Briles DE, Davie JM: Clonal dominance I. Restricted nature of the IgM antibody response to group A streptococcal carbohydrate in mice. *J Exp Med* 1975; 141:1291–1307.
43. Bloem A, Zenke G, Eichmann K, Emmrich F: Human immune response to group A streptococcal carbohydrate (A-CHO). II. Antigen-independent stimulation of IgM Anti-A-CHO production in purified B cells by a monoclonal anti-idiotypic antibody. *J Immunol* 1988; 140:277–282.
44. Rubinstein LJ, Stein KE: Murine immune response to the *Neisseria meningitidis* group C capsular polysaccharide. I. Ontogeny. *J Immunol* 1988; 141:4352–4356.
45. Diakun KR, Matta KL: Synthetic antigens as immunogens: Part III. Specificity of an anti–anti-idiotypic antibody to a carbohydrate tumor-associated antigen. *J Immunol* 1989; 142:2037–2040.
46. Misplon JA, Kindt RH, Reeves JP, Harvath L, Rubinstein LJ, Epstein SL: Induction of antigen-specific immunity by anti-idiotypic antibodies: Isotype expression in responses and potency of induction by monoclonal anti-idiotypes. In: Osterhaus, ADME, Urbain J, Uytdehaag FGCM, eds: *Proceedings, Idiotype Networks in Biology and Medicine Congress, Gennep. The Netherlands. 17–20 April 1989*. Amsterdam: Elsevier Science Publishers; 1990:51.

47. Bruck C, Co MS, Slaoui M, et al: Nucleic acid sequence of an internal image-bearing monoclonal anti-idiotype and its comparison to the sequence of the external antigen. *Proc Natl Acad Sci USA* 1986; 83:6578–6582.
48. Williams WV, London SD, Weiner DB, et al: Immune response to a molecularly defined internal image idiotype. *J Immunol* 1989; 142:4392–4400.

CHAPTER 5

Characterization of Idiotypic Reagents as Antigen Surrogates of a Human Tumor-Associated Antigen

Giovanna Viale and Antonio G. Siccardi

Introduction

A large number of monoclonal antibodies reactive with human tumors of different histological types have been characterized (1–3) and successfully employed in tumor diagnosis in vitro (serology and immunohistochemistry) (4, 5) and in vivo (tumor imaging) (6–8); moreover, their possible use in therapy is under extensive investigation (9, 10).

Increasingly there is evidence, however, that the molecules recognized by such operationally useful monoclonal antibodies are mostly tumor-associated, rather than tumor-specific antigens (11), that is, they are differentiation antigens that are also expressed on normal cells, at least during particular stages of the life cycle of these cells. It is thus not surprising that the human immune system does not usually react against these antigens, although other factors, such as blocking, nonspecific suppression, and poor expression of MHC antigens of class I and II on tumor cells could be involved.

Following Jerne's formulation of the network theory (12, 13), much attention has been focused on the possibility of exploiting the regulatory constraints exerted by idiotypic–anti-idiotypic (anti-Id) interactions among the elements of the immune system to control and modulate host immune responses to a large variety of haptens and protein and carbohydrate antigens (reviewed in 14, 15).

One of the most intriguing features of Jerne's view of the immune system is the postulated existence of anti-Id molecules complementary to paratope-associated idiotypic determinants. Such anti-Id antibodies (in Jerne's terminology, anti-idiotypes of class beta, or "internal images") must beat idiotypic structures immunologically equivalent to antigenic epitopes on external antigens. The collection of internal images (the idiotypic universe) therefore reflects, like a mirror, the antigenic universe of the outside world.

In many experimental systems, anti-Id antibodies of the internal image type have been successfully exploited to elicit specific immunity to viral, bacterial, and protozoal antigens (reviewed in 16). Both induction of neutralizing antibodies and/or protection against subsequent challenge with various infectious agents have been reported. Interestingly, the idiotypic vaccination approach has proved successful in an experimental system such as the chimpanzee for human hepatitis B virus immunization (17); furthermore, immunization trials in chimpanzees with anti-Id antibodies mimicking the CD4 receptor and neutralizing HIV infection of human T cells in vitro are now in progress (18).

In tumor immunology, anti-Id antibodies can modulate tumor growth both in vivo and in vitro, in experimental systems (19, 20) and in clinical trials (21, 22). Ab3 humoral immune responses (21–23), antitumor T effector cells, and/or idiotype-specific regulatory T cells (19, 20, 24, 25) have all been suggested as responsible for the modulation effects observed after anti-Id immunizations.

Internal images of tumor-associated antigens that may play a role in inducing anti-tumor immunity and anti-Id molecules that mimic both murine (26, 27) and human (28, 29) tumor markers have been described in different experimental systems.

Herlyn et al (22) have reported the effects of anti-Id antibody treatment in a clinical trial on 30 patients with advanced colorectal carcinoma. Six patients showed partial clinical remission and seven patients showed arrest of metastases: among these, four patients had received Ab2 alone and nine patients Ab2 in association with chemotherapy. In all the patients, immunization with Ab2 elicited highly specific antitumor antibody responses.

We have focused our attention on the idiotypic approach to produce anti-Id reagents able to substitute for the nominal antigen with the aims to (a) analyze the possible modulation of immune response to tumor and (b) exploit this strategy for the production of "second-generation" antitumor monoclonal antibodies.

MBr1 is a murine monoclonal antibody defining a saccharidic epitope (CaMBr1) of a human tissue-specific, tumorassociated globoside that is present on the mammary carcinoma cell line MCF-7 (30). This epitope, which was shown to be present on different molecular species (mammary glycoproteins of different molecular weight and mucins, in addition to glycolipids), is characteristic of normal and neoplastic mammary epithelial cells (30, 31); the same epitope (or a cross-reacting one) is carried on mucins from some ovarian cyst fluids. MBr1 is a useful immunocytological reagent (32) for the detection of metastatic foci in lymph node biopsies and in pleural effusions. In normal breast, the presence of CaMBr1 is related to the functional status of epithelial cells and is also associated with functional changes in the breast gland that are related to hormonal stimuli. Moreover, tumors expressing CaMBr1 are mostly well-differentiated carcinomas and include a high percentage of tumors positive for estrogen and transferrin receptors.

A retrospective study on 40 patients with breast carcinoma revealed the high predicting value of this marker, which has a stronger association with poor prognosis than other already validated parameters, such as clinical stage, hormone receptor levels, and nodal status (33). In a further study (34), it was shown that the poor prognosis could be related to CaMBr1 expression on glycoproteins rather than on glycolipids.

In this chapter we will review the development and characterization of two anti-Id monoclonal antibodies (35) that mimic the CaMBr1 epitope in both its antigenic and immunogenic properties and of an anti–anti-Id monoclonal antibody with antitumor activity (36).

Materials and Methods

Monoclonal antibodies. Ab1 (MBr1, IgM, k), Ab2 (A3B10, IgG1, k, and E6F7, IgM, k), and Ab3 (2G-3, IgM, k) monoclonal antibodies have been described (30, 31, 35, 36).

Generation of Anti-Id and Anti–Anti-Id Antibodies. BALB/c mice were immunized intraperitoneally with 50 μg of immunogen (MBr1-KLH or A3B10-KLH) in complete Freund's adjuvant (day 1). The second dose was given in incomplete Freund's adjuvant on day 30. On days 44 to 46, the mice received three intraperitoneal boosters in PBS (50 μg of MBr1-KLH or A3B10-KLH). Fusions were performed on day 47 by standard methods. Rabbits were immunized monthly with 100 μg of purified A3B10 (or E6F7) emulsified in complete Freund's adjuvant both by injection in the hind footpads and by intramuscular administration.

Affinity Purification of Rabbit Anti–Anti-Id Antibodies. Rabbit anti–anti-Id sera were first extensively absorbed on Sepharose derivatized with irrelevant monoclonal antibodies to remove anti-isotypic antibodies and were tested for residual activity in enzyme-linked immunosorbent assay (ELISA) on A3B10-coated plates. The sera were then absorbed on A3B10-Sepharose and the anti-Id antibodies were eluted with 0.1 M acetic acid, 0.5 M NaCl, immediately buffered with 1 M TRIS, and dialyzed against phosphate-buffered saline (PBS).

Screening of Ab2 and Ab3 Hybridomas. Initial screening of Ab2 hybridomas was done by hemagglutination of sheep red blood cells coupled with $CrCl_3$ (Baker Chemicals, Phillipsburg, New Jersey) to MBr1 or to irrelevant antibodies. Hybridomas producing hemagglutinating antibodies were expanded and characterized further in antigen competition assays as inhibitors of the binding of radiolabeled MBr1 to MCF-7 monolayers. The same assay was utilized as an antibody competition test to screen for Ab1-like activity within Ab3 hybridomas. Ovarian cyst mucins or tetrahydrofuran MCF-7 cell extracts (glycolipid extracts) (37) were dried in microtiter wells (5 μg/well) overnight at 37 °C under vacuum, saturated with 3% bovine serum albumin (BSA) in PBS, and used for binding and competition assays.

Idiotypic Characterization. Binding and competition assays on antibody-coated microtiter plates were carried out using either radioiodinated or biotinylated tracers. Anti–anti-Id activity of mouse sera was assessed in a sandwich assay, in which A3B10 was used both as a catcher and a biotinylated tracer.

Results

Screening of Anti-Id Monoclonal Antibodies. The supernatants of hybridomas obtained from two BALB/c mice immunized with MBr1-KLH were tested for anti-Id activity in hemagglutination assays with a panel of sheep red blood cells (uncoated, or derivatized with MBr1, or with irrelevant monoclonal IgM). Supernatants with anti-Id activity (50/400) were assayed as possible competitors of radioiodinated MBr1 binding to MCF-7 cells and two of them, A3B10 and E6F7, gave inhibition values higher than 70%. The corresponding hybridomas were recloned twice and expanded as ascitic tumors.

Purified antibodies were tested in competition assays of biotinylated MBr1 binding to MCF-7 cells, as well as to partially purified antigen (ovarian cyst mucins and glycolipid extracts). Table 5.1 shows that underivatized MBr1 inhibits to 50% the binding of biotinylated MBr1 to solid-phase MCF-7 cells at a concentration of about 4 μg/well. Under the same conditions, A3B10 is four times more efficient, as a competitor, than MBr1 itself; E6F7 was shown to have the same properties of A3B10, but to be 10 times less efficient. Analogous result were obtained on both ovarian cyst mucins and glycolipid extracts.

Reciprocal Competition Between A3B10 and E6F7. E6F7 was shown to inhibit the binding of radioiodinated A3B10 to solid phase MBr1 (35). Under the same conditions, however, A3B10 was 10-fold more efficient than E6F7. The difference between A3B10 and E6F7 reactivity in competition assays was confirmed by the finding that E6F7 did not appreciably inhibit the binding of biotinylated MBr1 to solid phase A3B10, whereas both liquid-phase MBr1 and A3B10 inhibited it completely. Taken together, these data indicated that both A3B10 and E6F7 are paratope-related anti-Id antibodies and that they

Table 5.1. Competition of biotinylated MBr1 binding to antigen by anti-Id reagents.

Competitor	MCF-7[a]	Source of antigen ovarian mucins[a]	Glycolipid extracts[b]
A3B10	1 μg[c]	1 μg	2.5 μg
E6F7	10 μg	4 μg	ND
NS1 ascites	>1 mg	>1 mg	>1 mg
MBr1	4 μg	6 μg	10 μg

[a] Competition on MCF-7 and on ovarian mucins was carried out with 1.3 μg/assay of biotinylated MBr1.
[b] Competition on glycolipid plates was carried out with 0.12 μg/assay of biotinylated MBr1.
[c] Dose of competitor giving 50% inhibition values.

Table 5.2. Characterization of anti–anti-Id mice and rabbit sera.

	Dilutions	Serum			
		MoAb3[a]	NMS	RaAb3[b]	NRS[c]
Anti–anti-Id activity	1:300	>2.0[d]	0.1[d]	1.6[e]	0.0[e]
Binding to MCF-7	1:300	1.0	0.2	1.1	0.1
Binding to glycolipids	1:50	0.6	0.2	0.9	0.3

[a] Samples collected before fusion.
[b] Rabbit serum at 4 months from the beginning of immunizations, absorbed on Sepharose derivatized with irrelevant monoclonal antibodies.
[c] Preimmune serum from the same rabbit.
[d] Determined in the sandwich assay described in Methods.
[e] Determined in direct binding assay on A3B10-coated plates.

interact with MBr1 with different affinities, that is, through nonidentical idiotopic structures.

Induction of Ab3 Responses in Mice and Rabbits. We have analyzed the immunogenic properties of the two reagents on the assumption that true *internal images* would act as antigen surrogates and elicit Ab1-like antibodies within Ab3 responses in mice and, more significantly, in animals of a different species. Ab3 responses were raised in mice and rabbits with KLH conjugates of A3B10 and E6F7. Anti–anti-Id activity was tested in a sandwich assay for mouse sera and in direct binding assays for rabbit sera. As shown in Table 5.2, on immunization with A3B10, high titer anti–anti-Id antibodies are elicited in both mice (MoAb3) and rabbits (RaAb3). Ab3 sera were also tested, in ELISA, for the presence of antibodies with Ab1-like anti-tumor activity. A3B10 induced anti-MCF-7 responses in both mice and rabbits; optimal responses were obtained, in mice, after 2 months and remained stable in the following months; in rabbits, optimal responses were obtained after 4 months. E6F7 induced stable responses in rabbits, although only transient responses were obtained in mice. These same sera were tested in binding assay on partially purified antigen (glycolipid extracts).

A partial idiotypic characterization of Ab3 sera was carried out by competition assays. Mouse Ab3 serum and affinity-purified rabbit Ab3 antibodies efficiently competed with the binding of MBr1 to solid phase A3B10 (Fig. 5.1). Conversely, MBr1 did not appreciably inhibit the binding of affinity-purified rabbit Ab3 to solid phase A3B10, which indicated that rabbit Ab3 antibodies contain only a minor component with Ab1-like properties (data not shown).

Characterization of Ab1-Like Antibodies Within Ab3 Responses. To analyze the fine specificity of Ab3 sera better, their pattern of reactivity was compared with that of MBr1 on normal and neoplastic mammary tissues by means of immunohistochemistry (Fig. 5.2). Monoclonal antibody MBr1 and polyclonal mouse and rabbit Ab3 sera reacted with ductal cells from normal mammary

Fig. 5.1. Inhibition of binding of [^{125}I]MBr1 to A3B10-coated plates. Competitors were mouse Ab3 (○), normal mouse serum (□), normal rabbit serum (●) at threefold dilutions ranging from 1:150 to 1:4500; MBr1(■) and purified anti–anti-Id rabbit antibodies (△) were used as competitors at threefold dilutions ranging from 1 to 0.03 μg.

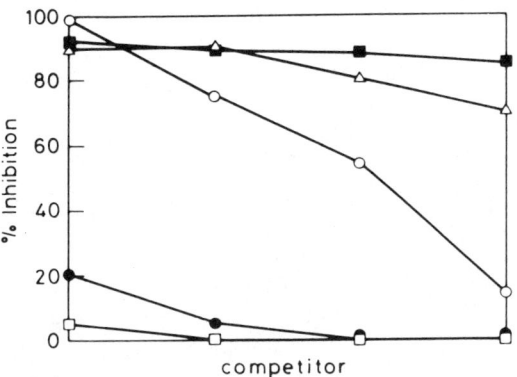

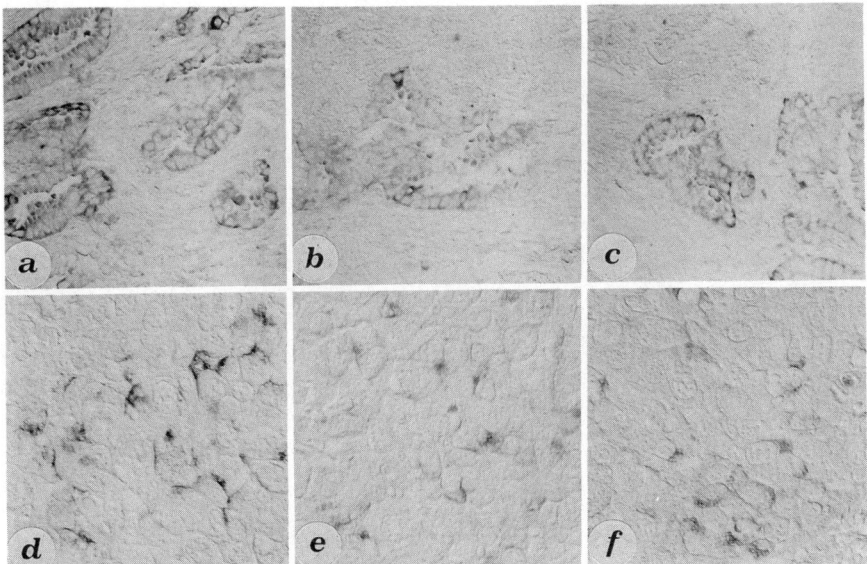

Fig. 5.2. Nonneoplastic breast (a, b, c) and ductal carcinoma (d, e, f) sections from the same paraffin block. The neoplastic cells show cytoplasmic immunoreactivity. a, d: Mbr1; b,e: mouse Ab3; c,f: rabbit Ab3. Antisera dilutions were 1/200. MBr1 was used at a concentration of 10 μg/ml. Immunoperoxidase, 4-Cl-1-naphthol substrate (a, b, c × 250; d, e, f × 400).

glands and ductal neoplasias. The pattern of reactivity of Ab3 sera on both normal and neoplastic tissues was superimposable on that obtained with MBr1, which indicated the presence of an Ab1-like component. In nonneoplastic and in some neoplastic tissues from tubular carcinomas, the immunoreactivity was confined to the apex of the cytoplasmic membrane; cells from ductal carcinomas also had cytoplasmic immunoreactivity. Control experiments were performed by means of indirect immunofluorescence on cytospin preparations of cell suspensions from MCF-7 and other human cell lines (Malme-3M, melanoma; HT-29, colon carcinoma; OvCa-432, ovarian carcinoma).

Characterization of Anti–Anti-Id Hybridoma Supernatants. Hybridoma supernatants from a Balb/c mouse immunized with A3B10-KLH were screened in binding assays on MCF-7 cells monolayers; positive supernatants (180/1600) were further tested as competitors of the binding of radioiodinated MBr1 to the same tumor cells. Only three supernatants gave inhibition values higher than 50%. The hybridoma producing monoclonal antibody 2G-3, which exhibited all the properties reported for Ab3 sera in Table 5.2, was selected, cloned twice, and further characterized as a candidate Ab1 "true clone" elicited by an anti-Id immunization (36).

Idiotypic and Antigen Binding Activity of 2G-3. The reactivity pattern of 2G-3 was compared to that of MBr1 in a number of assays (36), as summarized in Table 5.3. The binding activity of biotinylated 2G-3 monoclonal antibody to both A3B10 and E6F7 anti-Id antibodies was similar to that of biotinylated MBr1 when the same reagents were used. Further evidence of anti-Id activity of 2G-3 was given by its capacity to inhibit the binding of A3B10 to solid phase MBr1 to an extent similar to that exhibited by MBr1.

Purified 2G-3 and MBr1 had similar binding properties on MCF-7 intact cells. The antigen specificity of 2G-3 was also tested in binding assays to purified glycolipids; here binding proved to be less efficient than that of MBr1 at the same concentration, which indicated either a lower affinity of 2G-3 or a slight difference in specificity. Reciprocal competition assays supported this difference, since 2G-3 did not appreciably compete with biotinylated MBr1 in

Table 5.3. Comparison of 2G-3 and MBr1 reactivity patterns.

	2G3	MBr1
Binding to		
A3B10	+	+
E6F7	+	+
MCF-7	+	+
Glycolipid extracts	+	+
Glycolipid extracts		
(25 mM NaIO$_4$ treated)	−	−
Reciprocal competition		
2G3	+	−
MBr1	+	+
Western blotting		
Glycoproteins	+ +	+
Glycolipids	−	+ +

binding to solid-phase glycolipid, whereas MBr1 efficiently (>50%) competed with biotinylated 2G-3 in binding to the same solid-phase glycolipid.

The carbohydrate nature of the antigenic determinant recognized by MBr1 has been previously defined, since metaperiodate oxidation of MCF-7 crude membrane preparations drastically affected MBr1 binding activity (31). We showed that the binding of 2G-3 on glycolipid extracts was affected by $NaIO_4$ treatment in a dose-dependent way and that the same effect was obtained with MBr1 in a parallel experiment. This result suggested that carbohydrate residues on glycolipid molecules are the antigenic determinants that are recognized by both MBr1 and 2G-3.

The fine specificity of the two antibodies was also studied by Western blot analysis of MCF-7 extracts (Fig. 5.3c). Both antibodies recognized a major protein band of 43 kDa and a minor band of 50 kDa, although 2G-3 reacted more strongly than MBr1. On the same sample, MBr1 also recognized, as expected (38), the glycolipid band that migrates to the bottom of the gel. However, 2G-3 failed to react with the glycolipid band in Western blot analysis. This result indicates that the reactivity of 2G-3 to purified glycolipids in ELISA assays might be due to contaminating glycoproteins present in the tetrahydrofuran cell extracts. Alternatively, the saturation of the solid-phase glycolipid with bovine serum albumin might mimic a glycoprotein configuration.

Reactivity of 2G-3 and MBr1 in Immunohistochemistry. Only slight differences in reactivity between 2G-3 and MBr1 were observed in paraffin-embedded, formalin-fixed normal tissues. The only qualitative difference was observed at the level of kidney distal tubular epithelia, which were positive with MBr1 and

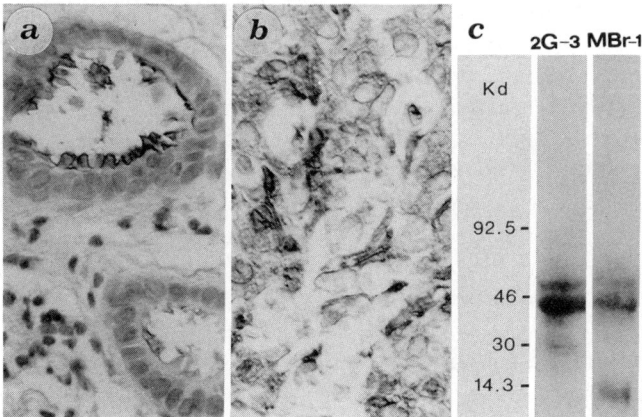

Fig. 5.3. Reactivity of 2G3 at the apex of ductal cells in (*a*) apocrine metaplasia and (*b*) in the cytoplasm of a ductal carcinoma. (*c*) Western blot analysis showing reactivity patterns of 2G3 and of MBr1 on Triton X-100 lysates of MCF-7 cells (*a* and *b* X 450; 3-3'-diaminobenzidine substrate; hematoxylin nuclear counterstain).

negative with 2G-3. The pattern of reactivity of the two antibodies on neoplastic tissues was superimposable, as shown in Figure 5.2 (a, d) and Figure 5.3 (a, b). The frequency of positive breast tumors did not significantly differ from that obtained with MBr1 (26/33 and 29/33, respectively) (39) and was higher for 2G-3 than for MBr1 (13/14 and 3/14, respectively) on a small series of ovarian carcinomas (36).

Conversely, immunohistochemical staining of acetone-fixed cryostatic sections of breast and ovary tumors demonstrated important quantitative and qualitative differences in reactivity. On a series of 40 adenocarcinomas (28 breast and 12 ovary), we found 12 cases that reacted with one of the two monoclonal antibodies but not with the other. Such a discrepancy is likely to reflect a differential expression of CaMBr1 on various glycoconjugates and might be of diagnostic interest in view of the reported correlation between CaMBr1 and poor prognosis (34).

Discussion

According to Jerne's theory (12), the immune system is regulated by a network of complementary molecular structures that allow the cells of the system to interact among themselves by exerting regulatory constraints. In the original formulation of the theory, the interacting structures are paratopes and idiotopes of antibody molecules and the complexity of the system is conceived to be such that the idiotopic repertoire would be as large as the epitopic "universe" and essentially overlap it. It follows that for any epitope of the external world there is an idiotype, within the immune system, that represents its immunological equivalent or *internal image*.

Animals immunized with Ab1 antibodies produce anti-Id Ab2 immunoglobulins that react with Ab1 molecules in either of two ways: (a) as "antibodies," binding through their paratope to Ab1 idiotopes (Ab2-alpha and Ab2-gamma anti-Id), or (b) as "antigens," bearing an idiotope complementary to the paratope of Ab1 (Ab2-beta). Ab2-beta molecules, which bear the internal image of the nominal antigenic epitope recognized by Ab1, will be capable of inducing, if used as immunogens in animals of the same or of different species, Ab3 responses containing antibodies with reactivity similar to or identical with that of Ab1 (Ab1 like). Indeed, it has been shown (40) that, in mice, the major outcome of immunization with an Ab2 beta is the activation of "true" Ab1 clones.

In this chapter, we have reviewed the cloning and characterization of two monoclonal Ab2-beta anti-Id antibodies and characterized both polyclonal and monoclonal Ab3 responses induced upon immunization with one of the Ab2 beta anti-Id antibodies. A3B10 and E6F7 are two paratope-related monoclonal anti-Id molecules induced against MBr1, a murine monoclonal antibody defining a saccharidic epitope (CaMBr1) of a human, tissue-specific, tumor-associated globoside, which is present on cells of the mammary carcinoma-derived human cell line MCF-7. The same epitope is shared by glycoproteins that are present on normal and neoplastic mammary epithelial cells and by mucins from some ovarian cyst fluids.

In order to be defined as *internal images*, the two anti-Id monoclonal antibodies were shown to meet both immunological (inhibition of binding of Ab1 to nominal antigen) and biological (ability to behave as immunogens across species barrier) criteria. Indeed, both A3B10 and E6F7 (a) compete with MBr1 binding to nominal antigen both on tumor cells and in a soluble form and (b) induce anti-tumor antibodies in mice and rabbits.

The bulk of our data leads to the conclusion that the three-dimensional structures carried by the two internal images anti-Id monoclonal antibodies show different levels of homology to the original antigen, with A3B10 always appearing to be the better image of CaMBr1. Such a difference in affinity indicates the nonidentity of the idiotopic structures carried by the two antibodies.

A further step in the idiotypic chain of immunization led us to the induction of both polyclonal and monoclonal anti–anti-Id molecules. Upon immunization with an anti-Id of the internal image type, antibodies with anti-tumor activity were elicited in polyclonal responses, and eventually a monoclonal anti–anti-Id hybridoma was isolated with binding properties comparable to those exhibited by MBr1. Preliminary data, however, suggest that, although monoclonal Ab1 and Ab3 react with the same saccharidic epitope, the fine specificity of MBr1 and that of 2G-3 are not identical. MBr1 was shown to be highly specific for glycolipid molecules, whereas 2G-3 preferentially binds glycoproteins.

A possible explanation for the preferential binding of the two reagents could reside in the chemical nature of the immunogens used to raise the two monoclonal antibodies (glycolipid versus protein internal image). Our results suggest that the molecular environment of antigen presentation influences the outcome of immunization and the activation of specific clones.

Our studies further support the use of anti-Id monoclonal antibodies as antigen surrogates to elicit antitumor responses and introduce the notion that Ab1-like molecules, endowed with slightly different specificities, can contribute to the fine characterization of the nominal antigenic epitope.

Acknowledgments. This work was supported by grants of Consiglio Nazionale delle Ricerche, Special Project "Oncologia" (86.00609.44, 87.011482.44 and 88.00775.44). We are grateful to Drs. Roberto Buffa and Flavio Leoni for contributing unpublished materials and to Ms. Silvana Spedalini for secretarial assistance.

References

1. Koprowski H, Herlyn M: Human tumor antigens. In: Wahren B, Holm G, Hammarstrom S, Perlmann P, eds: *Molecular Biology of Tumor Cells.* New York: Raven Press; 1985: P 123.

2. Hellström KE, Hellström I, Brown JP, Larson SM, Nepom GT, Carrasquillo JA: Genes and antigens in cancer cells: The monoclonal antibody approach. In: Eckhardt S, Holzner JH, Nagel GA, eds: *Contributions of Oncology*. Basel: Karger Publications; 1984.
3. Lloyd KO: Human tumor antigens: Detection and characterization with monoclonal antibodies. In: Herberman RB, ed: *Basic and Clinical Tumor Immunology*. Boston: Martinus Nijhoff Publishers; 1983.
4. Schlom J: Basic principles and applications of monoclonal antibodies in the management of carcinomas: The Richard and Linda Rosenthal Foundation Award lecture. *Cancer Res* 1986; 46:3225.
5. Bast RC, Klug TL, John E, et al: A radioimmunoassay using a monoclonal antibody to monitor the course of epithelial ovarian cancer. *N Engl J Med* 1983; 309:883.
6. Keenan AM, Harbert JC, Larson JM: Monoclonal antinbodies in nuclear medicine. *J Nucl Med* 1985; 26:531.
7. Siccardi AG, Buraggi GL, Callegaro L, et al: Multicenter study of immunoscintigraphy with radiolabeled monoclonal antibodies in patients with melanoma. *Cancer Res* 1986; 46:4817.
8. Siccardi AG, Buraggi GL, Callegaro L, et al: Immunoscintigraphy of adenocarcinomas by means of radiolabeled F(ab')$_2$ fragments of an anticarcinoembryonic antigen monoclonal antibody: A multicenter study. *Cancer Res* 1989; 49:3095.
9. Dillman RO, Royston I: Applications of monoclonal antibodies in cancer therapy. *Br Med Bull* 1984; 40:240.
10. DeLand FH, Goldenberg DM: Diagnosis and treatment of neoplasms with radionuclide-labelled antibodies. *Sem Nucl Med* 1985; 15:2.
11. Zalcberg JR, McKenzie IFC: Tumor-associated antigens. *J Clin Oncol* 1985; 3:876.
12. Jerne NK: Toward a network theory of the immune system. *Ann Inst Pasteur Immunol* 1974; 125c:373.
13. Jerne NK: Idiotypic networks and other preconceived ideas. *Immunol Rev* 1984; 79:5.
14. Bona CA, Kohler H: Immune networks. International Conference on "Immune Networks". *Ann NY Acad Sci* 1983; 418:1.
15. Rodkey LS: Autoregulation of immune responses via idiotype network interactions. *Microbiol Rev* 1980; 44:631.
16. Sacks DL, Kelsoc JH, Sachs DH: Induction of immune responses with anti-idiotypic antibodies: Implications for the induction of protective immunity. *Springer Semin Immunopathol* 1983; 6:79.
17. Kennedy RC, Eichberg JW, Lanford RE, Dreesman GR: Anti-idiotypic antibody vaccine for type B viral hepatitis in chimpanzees. *Science* 1986; 232:220.
18. Chanh TC, Dreesman GR, Kennedy RC: Monoclonal anti-idiotypic antibody mimics the CD4 receptor and binds human immunodeficiency virus. *Proc Natl Acad Sci USA* 1987; 84:3891.
19. Lee VK, Hellström KE, Nepom GT: Idiotypic interactions in immune responses to tumor-associated antigens. *Biochim Biophys Acta* 1986; 865:127.
20. Kennedy RC, Zhou E, Lanford RE, Chanh TC, Bona CA: Possible role of anti-idiotypic antibodies in the induction of tumor immunity. *J Clin Invest* 1987; 80:1217.
21. Koprowski H, Herlyn D, Lubeck M, De Freitas E, Sears F: Human anti-idiotype antibodies in cancer patients: Is the modulation of the immune response beneficial for the patient? *Proc Natl Acad Sci USA* 1984; 81:216.

22. Herlyn D, Wettendorff M, Schmoll E, et al: Antiidiotype immunization of cancer patients: Modulation of the immune response. *Proc Natl Acad Sci USA* 1987; 84:8055.
23. Dunn PL, Johnson CA, Styles JM, Pease SS, Dean CJ: Vaccination with syngeneic monoclonal anti-idiotype protects against a tumor challenge. *Immunology* 1987; 60:181.
24. Raychaudhuri S, Saeki Y, Chen J, Iribe H, Fuji H, Kohler H: Tumor-specific idiotype vaccine. II. Analysis of the tumor-related network response induced by the tumor and by internal image antigens (Ab2-beta). *J Immunol* 1987; 139:271.
25. Lee VK, Harriott TG, Kuchroo VK, Halliday WJ, Hellström I, Hellström KE: Monoclonal anti-idiotypic antibodies related to a murine oncofetal bladder tumor antigen induce specific cell-mediated tumor immunity. *Proc Natl Acad Sci USA* 1985; 82:6286.
26. Forstrom JW, Nelson KA, Nepom GT, Hellström I, Hellström KE: Immunization to a syngeneic sarcoma by a monoclonal auto-anti-idiotypic antibody. *Nature* 1983; 303:627.
27. Raychaudhuri S, Saeki Y, Fuji H, Kohler H: Tumor-specific idiotype vaccines. *J Immunol* 1986; 137:1743.
28. Herlyn D, Ross HA, Koprowski H: Anti-idiotypic antibodies bear the internal image of a human tumor antigen. *Science* 1986; 232:100.
29. Nepom GT, Nelson KA, Holbeck SL, Hellström I, Hellström KE: Induction of immunity to a human tumor marker by in vivo administration of anti-idiotypic antibodies in mice. *Proc Natl Acad Sci USA* 1984; 81:2864.
30. Ménard S, Tagliabue E, Canevari S, Fossati G, Colnaghi MI: Generation of monoclonal antibodies reacting with normal and cancer cells of human breast. *Cancer Res* 1983; 43:1295.
31. Canevari S, Fossati G, Balsari A, Sonnino S, Colnaghi MI: Immunochemical analysis of the determinant recognized by a monoclonal antibody (MBr1) which specifically binds to human mammary epithelial cells. *Cancer Res* 1983; 43:1301.
32. Ménard S, Rilke F, Della Torre G, et al: Sensitivity enhancement of the cytologic detection of cancer cells in effusions by monoclonal antibodies. *Am J Clin Pathol* 1985; 83:571.
33. Colnaghi MI, Ménard S, Da Dalt MG, et al: A multiparametric study by monoclonal antibodies in breast cancer. In: Ceriani R, ed: *Immunological Approaches to the Diagnosis and the Therapy of Breast Cancer*. New York: Plenum Press; 1987:p 21.
34. Miotti S, Leoni F, Canevari S, Sonnino S, Colnaghi MI: Immunoblotting detection of carbohydrate epitopes in glycolipids and glycoproteins of tumoral origin. In: Oettgen HF, ed: *Gangliosides and Cancer*. New York: VCH Publisher; 1989: p 167.
35. Viale G, Grassi F, Pelagi M, et al: Anti-human tumor antibodies induced in mice and rabbits by "internal image" anti-idiotypic monoclonal immunoglobulins. *J Immunol* 1987; 139:4250.
36. Viale G, Flamini G, Grassi F, et al: Idiotypic replica of an anti-human tumor-associated antigen monoclonal antibody: Analysis of monoclonal Ab1 and Ab3 fine specificity. *J Immunol* 1989; 143:4338.
37. Tettamanti G, Bonali F, Marchesini S, Zambotti V: A new procedure for the extraction, purification and fractionation of brain gangliosides. *Biochim Biophys Acta* 1973; 296:160.

38. Leoni F, Miotti S, Canevari S, Sonnino S, Ripamonti M, Colnaghi MI: Carbohydrate epitope defined by an antitumor monoclonal antibody detected on glycoproteins and a glycolipid by immunoblotting. *Hybridoma* 1986; 5:289.
39. Mariani-Costantini R, Barbanti P, Colnaghi MI, Ménard S, Clemente C, Rilke F: Reactivity of a monoclonal antibody with tissues and tumors from the human breast. *Am J Pathol* 1984; 115:47.
40. Rubinstein LJ, Goldberg B, Hiernaux J, Sten KE, Bona CA: Idiotype-antiidiotype regulation. V. The requirement for immunization with antigen or monoclonal anti-idiotypic antibody for the activation of beta2→6 and beta2→1 polyfructosanreactive clones in Balb/c mice treated at birth with minute amounts of anti-A48 idiotype antibodies. *J Exp Med* 1983; 158: 1129.

CHAPTER 6

Anti-Idiotypic Tumor Vaccines

Heinz Köhler and Sybille Müller

Introduction

Host defense against cancer is known to be usually insufficient to halt the growth and the dissemination of the tumor. This may be due to ineffective stimulation of the lymphocyte clones, which have the potential to mount a response to tumor-associated antigens (TAAs) (1) or due to suppression of cytotoxic T lymphocyte activity by progressive tumor growth (2). A common explanation for the absence of antitumor immunity is that the immune system has been tolerized by the tumor antigen (3–6); however, the exact mechanisms of suppression of immunity against TAAs are not known. Since stimulation of dormant antitumor-specific immunity with nominal tumor antigens has had only limited success, an alternative approach to immunize experimental animals or cancer patients with "internal image" (Ab2β or "network" antigens) (7), that is, anti-idiotypic (anti-Id or Ab2) antibodies, has been applied. This method may be effective in breaking tumor antigen-induced tolerance by presenting the critical TAA epitope in a different molecular environment to the tolerized host (8) and may thus break tolerance in tumor-specific immunity. Furthermore, the anti-Id antibodies, the so-called *network antigens* (7), are the most promising because they are not dependent on genetic idiotypic availability. Consequently, network antigens for a given tumor species would be effective in outbred populations.

Anti-Id antibodies (Ab2s) are directed against idiotypes on antibodies (Ab1) directed to an external antigen. The idiotypes may or may not overlap with the antigen-binding site (paratope) of Ab1. Among several types of anti-Id antibodies, some "mimic" the structures of external antigens and thereby induce an immune response similar to that induced by the original external antigen. The biological effectiveness of anti-Id antibodies in comparison to the external antigen is controlled and established by the affinity of the anti-Id antibody to the pre-existing idiotype immunoglobulin (Ig) receptor and the regulatory segments in the immune network, which decidedly influence the outcome of immune stimulation or suppression. By definition, *internal image* (Ab2β or *network antigen*) binding to an idiotype located on an antibody (Ab1) specific for the external antigen should be inhibited by the nominal (external) antigen. Using these anti-Id antibodies as antigen surrogates to induce specific responses, the utility of anti-Id antibodies as a vaccine for different pathogens was explored in several studies (for review see 9–11). It has been shown that anti-Id antibodies are not only useful for providing protective immunity as surrogate antigens against viral and bacterial infection (11), but they can also induce the expression of "silent" clones, which are included in the B cell repertoire but cannot be activated by the nominal antigen, and which act to break tolerance to antigen (12, 13). In this chapter, results from studies of the anti-Id approach in an animal tumor model, the DBA/2 lymphoma L1210, along with studies demonstrating an anti-Id–induced host response in patients with malignancies will be discussed.

Köhler and colleagues entry into tumor immunotherapy using the anti-Id approach several years ago was motivated by several observations: For instance, anti-Id antibodies have been applied in the treatment of B cell lymphoma in passive immunotherapy, but in many cases, success was limited due to spontaneous generation of somatic variants (14–20). Furthermore, Kennedy et al (21) used an anti-Id antibody to induce immunity to SV-40–transformed cells. Mice vaccinated with this anti-Id antibody exhibited prolonged survival after tumor transfer.

Tumor-Specific Transplantation Antigens and Tumor-Associated Differentiation Antigens Mimicking Anti-Id Antibodies

Antitumor immunity induced by tumor-specific transplantation antigens (TSTA) and tumor-associated differentiation antigens (TADAs) mimicking Ab2 has been observed by Hellström, Hellström, and others in a series of experiments over the last few years. Nelson et al (22) generated mouse monoclonal antibodies (MAbs) 4.72 and 5.96 against chemically induced sarcomas MCA-1790 and MCA-1511 in BALB/c mice that expressed TSTA. Hybridoma antibodies were tested for their ability to prime syngeneic mice for tumor-specific, delayed-type hypersensitivity (DTH). The MAbs 4.72 and 5.26 could induce DTH to tumor cells when injected into syngeneic mice for sensitization 5 days prior to challenge with tumor cells in the footpad. The MAbs 4.72 and 5.96 were designated autoanti-idiotypic antibodies, since they acted as surrogate antigens of MCA-1490 or MCA-1511 sarcoma cells and induced TSTA-specific DTH only in syngeneic mice (22).

Injection of the lymphocytes prevented the outgrowth of transplanted MCA-1490 sarcoma cells. Finally, it was demonstrated (22) that MAb 4.72 could specifically bind to a previously isolated tumor antigen-specific suppressor factor related to MCA-1490. This result suggested that suppressor factors unique for the TSTA of mouse sarcomas may express idiotopes that can be identified by the proper Ab2.

The findings obtained by studying Ab2 in relation to TSTA of murine tumors indicate that anti-Id vaccines can induce cell-mediated tumor immunity, which can lead to destruction of tumor cells in vivo.

Most human tumor antigens so far identified are TADAs, which are also present at low levels in normal tissues. It is therefore important to know whether Ab2 can be used to induce an immune response to TADA in its naive hosts.

The TADA p175, identified by a rat MAb, 6.10, is strongly expressed in mouse bladder carcinomas and some fetal cells and is scattered in epithelial cells from adult mice (23).

Lee et al (13) described two mouse MAbs, 21D9 and 4310, directed against a rat MAb that recognizes TADA p175. Both induced DTH to mouse bladder carcinoma in syngeneic BALB/c mice, whereas MAb 5.96, generated in relation to sarcoma MCA-1511 (see preceding section), was used as a control and only induced DTH to MCA-1511. The findings obtained in the mouse bladder carcinoma system showed that MAb2 can be used to induce cell-mediated tumor immunity to a TADA. This is important in view of the fact that the TADA studied was a "self" antigen expressed at low levels by certain normal, fully differentiated mammalian host cells.

Nepom et al (24) performed studies to determine whether an Ab2 that mimics TADA, p97, would induce both cell-mediated immunity and a humoral Ab1-like Ab3 response; p97 is a human melanoma-associated antigen (24). There workers made polyclonal Ab2s by immunizing rabbits against a mouse anti-p97 MAb, 8.2. After absorption and in vitro characterization for anti-Id activity, the rabbit sera were injected into mice, in which they induced both antibodies that precipitated p97 and cell-mediated immunity detected as DTH to human tumor cells expressing p97.

Subsequently, monoclonal Ab2s related to p97 were obtained (25). Some of these MAb2s induced an Ab1-like Ab3 response when given to either syngeneic or allogeneic mice. The fact that there was also a response in allogeneic mice implies that it was not allotype restricted (26, 27) and suggested that the MAb2 acted as an internal image of p97 (25). However, no cell-mediated immunity, measured as DTH, could be induced to p97. Furthermore, injection of MAb2 in mice did not protect against the outgrowth of transplanted mouse melanoma cells that expressed p97 after transfection with the p97 gene. Protection, however, was observed in mice that received a recombinant vaccinia virus vaccine containing the p97 genome (28). A p97 subunit vaccine, which was tested in parallel, induced high titers of humoral antibodies but no cell-mediated immunity to p97, and it failed to induce the rejection of p97-positive mouse melanoma cells (28).

Idiotype Vaccines Against Human T Cell Acute Lymphoblastic Leukemia

To explore the feasibility of the anti-Id vaccine approach for human tumor, Bhattacharya-Chatterjee et al (29) have begun studies with human T cell leukemia. Human T cell acute lymphoblastic leukemia (T-ALL), a common form of childhood leukemia, is associated with poor prognosis and a high rate of relapse (30–34). At present it is very difficult to cure T-ALL patients with relapsed disease (31, 33). It was their aim to prepare anti-Id (Ab2) antibodies against T-ALL to be used as an antigen substitute for the induction of therapeutic immunity. A previously described murine monoclonal antitumor antibody (Ab1), termed SN2, defines the protein moiety of a unique human T cell leukemia-associated cell surface glycoprotein, gp37, with an approximate molecular weight of 37,000 Da (35, 36). This SN2 is of the IgG1-κ subclass and can induce specific cytotoxicity against human T leukemia cells when conjugated with the A chain of ricin (37, 38). No significant antigenic modulation was observed when T leukemia cells were reacted in vitro with SN2. Generated syngeneic monoclonal mouse anti-Id antibodies against SN2 induced antitumor antibodies specific for the TAA defined by SN2.

The Ab2s to SN2 were screened on the basis of their binding to the F(ab')$_2$ fragments of SN2 and not to the F(ab')$_2$ fragments of pooled normal BALB/c mouse sera IgG1 or to an unrelated BALB/c MAb of the same isotype. Fifteen Ab2s, obtained from two fusions, were specific for the SN2 idiotope and not specific for isotype or allotype determinants. To find out whether these Ab2s were directed against the paratope of SN2, the binding of radiolabeled SN2 to leukemic MOLT-4 and JM cells, which contain gp37 as a surface constituent, was studied in the presence of these anti-Id antibodies. Clone 4EA2 inhibited the binding 100% at a concentration of 50 ng and 4DC6 inhibited the binding 90% at a concentration of 250 ng. A third clone, 4DD6, gave about 50% inhibition. The inhibition of SN2 binding to insolubilized MOLT-4 antigen or cell membrane preparation was similar. The binding of SN2 (Ab1) to 4EA2 and 4DC6 was also inhibited by a semipurified preparation of gp37 antigen. These results demonstrated that at least two of the anti-Id antibodies were binding either at or near the binding site idiotope of SN2. Next, the purified Ab2 was used to immunize syngeneic mice to induce antibody binding to MOLT-4 cells or gp37. Sera from mice immunized with 4EA2 and with 4DC6 coupled to keyhole limpet hemocyanin (KLH) contained antibodies that bound to semipurified gp37 antigen and molt-4 cells. Immune sera inhibited the binding of iodinated Ab2 to Ab1, which indicated that an anti–anti-Id antibody (Ab3) in mice shared idiotopes with Ab1 (SN2). Also, the binding of iodinated Ab2 to Ab1 was inhibited by rabbit antisera specific for gp37.

To characterize these anti-idiotopes further, they were used to immunize mice and rabbits (39). Several murine anti–anti-Id MAbs (Ab3s), mostly of IgM-κ isotype, were obtained. The MAb3 and sera from rabbits immunized with Ab2 contained antibodies that bound to gp37 antigen and leukemic MOLT-4 and JM cells. Also, MAb3 and immune sera from rabbits competed with Ab1 for binding to MOLT-4 cells. They inhibited the binding of iodinated Ab1 to Ab2, which indicated that Ab3 in mice and rabbits shares idiotopes with Ab1 (SN2). Furthermore, both the murine MAb3 and rabbit polyclonal Ab3 immunoprecipitated the same gp37 antigen as SN2 (Ab1). The production of antigen-specific Ab3 (Ab1') in mice and rabbits in the absence of any exposure to gp37 indicates that these Ab2s may indeed carry the internal image of the gp37 antigen. Collectively, these data suggested that anti-Id antibodies 4EA2 and 4DC6 carrying the internal image of gp37 might be candidates for idiotype vaccines against human T cell leukemia.

Ab2 in Patients with Solid Cancer

From the data to be discussed in the following section it was suggested that cancer patients would develop Ab2 when receiving antitumor Ab1 in passive immunotherapy. It was also suggested that an Ab1-like Ab3 response can be induced in cancer patients who have previously received Ab2 as an internal image or network antigen of the TAA. Both mechanisms of inducing Ab2 and Ab3 may produce a clinically beneficial patient response.

For instance, Koprowski et al (40) have treated human colon carcinoma patients with a mouse MAb, 17-1A, which primarily reacted with human gastrointestinal and pancreatic carcinoma. Several of the patients developed Ab2s to MAb 17-1A, and many of these patients underwent partial or complete tumor regression. The evidence for a clinical response was first observed several weeks after treatment with MAb1; however, no responses were seen in patients who did not produce Ab2. It was concluded that the beneficial clinical effects were not caused by killing of cancer cells induced by MAb 17-1A but rather by a host response induced to the tumor initiated by MAb 17-1A and subsequently by Ab2. When the B lymphocytes from the responding patient were cultured with Ab2 derived from this serum, an Ab3 that expressed tumor specificity resembling that of 17-1A was produced (41). Subsequently a pilot trial was performed on patients with metastatic carcinoma who received a polyclonal Ab2 raised in goats. Several of the treated patients had Ab3 in their sera with specificity similar to 17-1A, and some of these patients appeared to benefit clinically (42).

Kusame et al (43) made monoclonal mouse Ab2s to a MAb that recognized a proteoglycan antigen on human melanoma, and generated several Ab2s that functioned as internal images when tested in vitro. Such Ab2s were injected into patients with metastatic melanoma, some of whom developed Ab1-like Ab3. There was evidence that suggested a more favorable clinical course in a few of these patients.

Passive and active idiotype vaccine therapy in humans is presently being performed with mouse monoclonal or goat polyclonal antibodies as described above, as well as by other groups (19, 44, 45). Concerns have been raised about the use of foreign Igs to which human antibody responses would occur. Indeed, such human anti-mouse antibody (HAMA) responses in mouse Ig-treated patients have been observed (46, 47). Although these HAMA responses remain a matter of concern, the fact that the human immune system recognizes Fc-related epitopes on murine antibodies might be beneficial in some cases, when considered along with the concept of "intra(antigen) molecular help" to be discussed later.

The L1210 Murine Tumor System

Since the evidence of antitumor immunity induction in cancer patients by Ab2 therapy is preliminary, the mechanisms of regulation of the antitumor response induced by Ab2 remain to be elucidated in parallel in an experimental tumor system.

To explore the specificity and biological effects of active anti-Id immunotherapy in tumors, Raychaudhuri and coworkers (48-50) performed a series of experiments in an animal tumor model, the DBA/2 lymphoma L1210. This tumor has been well characterized, and several sublines have been obtained that differed in antigenicity and other biological properties at the beginning of these studies (48, 49). The L1210/GZL cells used were derived from an antileukemic, drug-resistant subline of DBA/2 lymphoma L1210. The L1210/GZL tumor exhibits an elevated expression of a TAA cross-reactive with the mouse mammary tumor virus (MMTV) envelope glycoprotein 52 (gp52). Several syngeneic MAbs against L1210-associated antigens have been produced (50). In the first study (48), the tumor-specific immune response induced by irradiated tumor cells (L1210/GZL) and anti-Id antibodies was analyzed. Monoclonal anti-Id antibodies (Ab2s) were made against the paratope of a monoclonal antitumor antibody (Ab1; 11C1) that recognizes a shared determinant of gp52 of MMTV and the TAA L1210/GZL.

Hybridomas expressing the internal image of gp52 (Ab2) were screened for inhibition of the binding of Ab1 to Ab2 by TAA L1210/GZL. Mice sensitized with irradiated L1210/GZL cells produced specific DTH when challenged into the footpad with several Ab2 hybridoma. In turn, Ab2 hybridoma supernatant or purified Ab2 induced a specific DTH reaction in mice when first sensitized with Ab2 and then challenged with L1210/GZL tumor cells. Fluorescence-activated cell sorter analysis demonstrated that fluorescence staining of L1210/GZL cells by 11C1 (Ab1) can be completely inhibited with preabsorption on Ab2 hybridoma cells. Mice immunized with one of the Ab2s coupled to KLH; the monoclonal anti-Id antibodies, 2F10, raised against 11C1 (Ab1); and 3A4, raised against 2B2 (Ab1), contained Ab1-like Ab3 antibodies able to bind MMTV in their sera. More important, in mice immunized with 2F10-KLH (Ab2), a significant inhibition of L1210/GZL tumor growth and subsequent prolonged survival could be observed (49). Mice preimmunized with 3A4 or D11 anti-Id antibodies did not reject the L1210/GZL tumor.

In order to explain why 2F10 anti-Id, but not 3A4 anti-Id, antibodies provided protection against tumor growth on a cellular level, the antitumor immunity induced by conventional and idiotypic immunization was compared at the level of cytotoxic T lymphocytes (CTL) (49). The results showed that immunization with both irradiated tumor cells as nominal antigen or monoclonal anti-Id antibodies (Ab2) induced CTL immunity specific for the L1210/GZL tumor. Cytotoxic T cells, generated by immunization with irradiated tumor cells, lysed 2F10- and 3A4-specific hybridoma cells. The frequency of tumor-reactive CTLs was found to be similar in mice immunized with Ab2 or irradiated tumor cells when examined at the precursor level. The depletion of a L3T4 + T cell population from 2F10 immune mice, which have protective immunity against the growth of L1210/GZL tumor cells, was found to increase the effectiveness of transferred T cells to inhibit tumor growth. The inability of 3A4 to induce antitumor immunity could, however, be correlated with the presence of a population of Lyt-2 + regulatory T cells. These results demonstrated that selection of binding site-related Ab2s may not be a sufficient criterion for the development of an idiotypic vaccine. Therefore, the frequency of T helper (Th) cells that recognize biologically active Ab2s in the L1210/GZL experimental tumor system was determined and analyzed.

Idiotype-recognizing Th cells induced either by anti-idiotypes, irradiated tumor cell antigen, or progressively growing tumor were studied in order to see if there were qualitative differences in the expression and frequencies of idiotype-recognizing T cells induced by tumor antigen and by Ab2 made against Ab1 (50). Immunization with both tumor network antigen and tumor cells could induce Th cells that recognized idiotopes present on the anti-Id antibody (internal image). However, idiotype-recognizing Th cells induced by live tumor appeared to have a preference for the idiotope that did not induce protective immunity. The reactivity patterns of idiotype-recognizing T cells obtained from 2F10 and from irradiated tumor immunized mice were similar in nature in the sense that Lyt-2 − T cells, that is, purified T cytotoxic/suppressor cell (Lyt-2 +) depleted T cells, obtained from these immunized mice, responded to both 2F10 and 3A4 as antigen, although Lyt-2 − T cells from tumor-immunized mice responded better to 3A4 antigen. However, the idiotype-recognizing T cells obtained from 3A4-immunized mice showed a similar reactivity pattern to T cells isolated from mice during the early phase of tumor growth (within day 4 to 5 after the inoculation of 10^4 live tumor cells). Lyt-2 − T cells isolated from mice immunized with 3A4 and mice with early tumor growth responded only to 3A4 antigen.

To determine the specificity of Th cells, Raychaudhuri et al (50) have used a helper assay to measure the cooperation of idiotype-recognizing T cells. Therefore, the ability of idiotype-recognizing T cells to help trinitrophenol (TNP)–KLH-primed B cells to produce specific antibody to Ab2 coupled to TNP was studied. It is intriguing to note that the growing tumor during the early phases of tumor growth induced Th cells that were idiotypically similar to 3A4-induced T cells. The inability of Lyt-2 − T cells, isolated from 4- to 5-day-old tumor in mice, to react with 2F10-TNP was not due to the absence of 2F10 idiotype-recognizing T cells, because 2F10 idiotype-recognizing T cells were present when examined at the precursor cell level. The observation that immunization with irradiated tumor induced idiotype-recognizing T cells that were qualitatively similar to those induced by 2F10 but different from 3A4-induced Th cells supports the existence of an operative regulatory network.

Regulatory T Helper Cells

The biological meaning of the finding that protective and nonprotective Ab2s (2F10 and 3A4) induced qualitatively different Th cells must be discussed. It is unlikely that the difference in antibody production in the 2F10- and 3A4-immunized groups was due to the generation of allotype-specific regulatory mechanisms because both 2F10 and 3A4 bear the same allotype (IgH1-e) and isotype (IgG2a). Therefore, any allotype-induced regulatory mechanism would be expected to affect both 2F10- and 3A4-specific responses equally. The inability of 3A4-immunized T cells to react with TNP-primed B cells in the presence of 2F10-TNP antigen, despite the presence of 2F10-specific precursors, might have a different explanation: It is possible that private idiotopes on the 3A4 molecule are more immunogenic than the idiotypes that are shared between 2F10 and 3A4, and that immunization with 3A4 would only cause the induction of T cells specific for private idiotypes. A more interesting possibility would be the existence of a regulatory T cell network that only allows the expansion and expression of 3A4-specific Th cells by suppressing the expansion of 2F10-specific T cells. However, these mechanisms could not be precisely addressed with heterogeneous populations of T cells.

To determine whether the 2F10 idiotype-recognizing cells were the key to the induction of an effective immune network, T cell clones were used. T cell lines and subclones responding to 3A4 and 2F10 were generated and characterized in order to determine the role of idiotype-recognizing T cells in maintaining down-regulation or up-regulation of an antitumor immune response (52).

Characterization of T Cell Clones from 3A4-Induced T Cell Lines

As outlined earlier, 3A4 is a paratope-specific anti-Id MAb and induces cellular and humoral anti-TAA responses. Saeki et al (52) determined the specificity, major histocompatibility complex (MHC) restriction, and antigen-processing requirement of Th cell lines and clones that recognize an idiotypic determinant on 3A4. These 3A4-idiotype-specific T cell clones were phenotypically Th cells, which recognized idiotypic determinants in the context of MHC class II molecules under MHC restriction. However, the 3A4-idiotype-specific T cell clones responded only to 3A4 and not to TAA. Because these clones did not recognize TAA, their biological role in antitumor immunity could be as regulatory T cells involved in the idiotypic network regulation.

To determine what kinds of T cell populations respond in 3A4-induced T cell lines, the T cell line was cloned. Isolated T cell clones could be classified as at least three types according to their specificities. The first type was represented by the TCL1-3 clone, which proliferated in response to irradiated syngeneic spleen cells irrelevant to the presence of any additional antigen.

They might be classified as autoreactive T cells. The second type was the TCL1-4 clone, which responded to both 3A4 and 2F10, but not to the relevant negative control MAb, UPC10. However, 3A4 and 2F10 were identical isotypically and allotypically; UPC10 was isotypically identical but allotypically different (UPC10:Igh1-a), and all MAbs and κ L chains. Consequently, the second type of T cell clone should recognize a common determinant shared by 3A4 and 2F10. To determine which kinds of common determinants the second type of T cell clone recognized, the response to the Fc fragment of 3A4 was tested. All of the second type of T cell clones (TCL1-4, 2-9, 2-12, 6-7, 6-8) proliferated upon stimulation with the Fc fragment of 3A4. Accordingly, the second type of T cell clones were specific for the Igh1-e allotype. The third type of T cell clone, represented by the TCL8-23, showed a significant proliferative response to 3A4 antibody and its Fab fragment, but not to any other MAb (2F10, UPC10) nor to the Fc fragment of 3A4. Other T cell clones (1-7, 4-5, 7-8, 7-22) had a specificity similar to the TCL8-23 clone. Therefore, the third type of T cell clone might be specific for a private idiotypic determinant on the 3A4 antibody. To confirm this interpretation, the specificity of TCL8-23 was further analyzed by using a panel of MAbs. The TCL8-23 clone showed a specific proliferative response to 3A4 but not to MAb2s 2F10 and 3F3, MAb1 11C1, or unrelated MAb (UPC10 and R10.8) (52). These results reinforced the evidence that the TCL8-23 clone was a 3A4-idiotype-specific T cell.

By phenotypic analysis, these 3A4-idiotype-specific T cell clones were determined to be Th cells (Thy 1.2+, L3T4+, Lyt2−). Th cells require antigen presentation by antigen-presenting cells (APCs; macrophages, B cells, and dendritic cells) and recognize antigenic determinants in the context of MHC class II molecules (53). In the case of idiotype-recognizing T cells, the question of MHC restriction is controversial (54). Here it was shown that these 3A4-idiotype-specific T cell clones recognized the 3A4 idiotype in the context of the MHC class II molecule under MHC restriction. Furthermore, pretreatment of APC with chloroquine inhibited the proliferative response by these 3A4-idiotype-specific T cell clones. Those results provided strong evidence that T cells can recognize idiotypic determinants on Ig molecules in the context of MHC class II molecules under MHC restriction. In addition, it was shown that antigen processing may be required, as with any conventional soluble protein antigen. In other words, soluble Ig molecules might be internalized, degraded by APC, and re-expressed as idiotypic determinants on the surface of APC to be presented to T cells.

The presentation of idiotypic determinants to T cells on the surface of APC bears relevance to the induction of tumor regression in vivo by means of anti-Id or anti-Ig antibodies. Ochi et al (55) have targeted the antigen KLH to the sIg of a class II-positive B cell lymphoma specific for TNP using either TNP–KLH or KLH covalently bound to an anti-Id antibody. They showed that both forms of modified KLH were presented by the tumor cells. Such presentation resulted in the killing of the lymphoma cells by class II-restricted helper/cytotoxic T cells. Here a foreign antigen bound specifically to tumor cells selectively sensitized these cells for recognition and killing by MHC class II-restricted cytotoxic T cells. Tumor cells took up, processed, and presented soluble antigens and thereby target T cells to tumor cells. In this way, the immune response to the tumor was redirected to a conventional antigen. The advantage is, as Ochi et al (55) pointed out, that processed antigen can persist on tumor cells for extended periods of time and will therefore allow time for the tumor cells to be recognized by antigen-specific T cells. Lanzavecchia and colleagues (45) have recently shown that it is possible to use xenogenic MAbs as antigens for T cell targeting. This work is providing clues to the mechanism by which anti-Id or anti-Ig antibodies induce tumor regression. Lanzavecchia et al (45) have cloned T cells from patients injected with mouse anti-Ig antibodies reacting with B cell lymphoma cells. These T cells recognized processed mouse Ig on autologous APC in an MHC-restricted fashion and can be used to focus T cells onto B cell lymphoma cells that express class II molecules. A major fraction of the mouse Ig-specific T cell clones showed strong cytotoxic effects for the lymphoma cells, thereby indicating a possible role in antitumor killing in vivo.

In addition to the effect of anti-Id–mediated focusing of T cells to Ig-expressing tumor cells, the capacity of B cells to bind, process, and present anti-idiotypes to T cells may be the mechanism by which a specific immune response is induced with anti-idiotype vaccination. This mechanism would be expected to operate with xenogeneic anti-idiotype vaccines or after coupling to carriers (55). Th cells recognizing foreign Fc epitopes on the anti-Id antibody become primed, while B cells that recognize the TAA network antigen expressed by the Fab domain are expanded. The result of this T–B interaction of Fc-specific Th and Fab-specific B cells would induce secretion of TAA-specific antibodies.

Cooperation of Fc-Specific T Helper Cells with Anti-Idiotype–Primed B Cells to Produce Antitumor Antibodies

An in vitro system consisting of cloned allotype (Fc)-specific Th cells and B cells from anti-idiotype-primed mice was set up (56). When these cultures were stimulated with the corresponding anti-Id antibody, the T cells proliferated and the B cells produced antibodies that bound to tumor cells specifically. Thus, it appeared that the "intra(antigen) molecular help" provided by Th cells recognizing a constant region determinant cooperated with B cells recognizing a variable region determinant (idiotypic determinant), that is, internal images or network antigens. These results provided important information for the formulation of idiotype-based immunotherapy of human tumors.

Since T cell help is required for an effective antibody response induced by anti-Id antibodies mimicking a TAA, such help could come from T cells recognizing other epitopes on the anti-idiotype. As mentioned, this concept has recently been formulated by Lanzavecchia (57), who proposed exploring the immune system's own strategies for immunotherapy. This notion, originally made for cytotoxic T cells recognizing anti-Ig–coated tumor cells, can be extended to the induction of tumor-specific antibodies. Therefore, in the study discussed in the following section (56), cloned Th cells and the monoclonal anti-Id antibody 2F10 were used to investigate tumor-specific, anti-Id vaccination.

2F10, and anti-Id antibody derived fron an Igh1-e allotype mouse strain, which induces protection against the L1210/GZL DBA/2 tumor, as discussed earlier, was used to prime DBA/2 mice. An Fc (Igh1-e)-specific syngeneic Th clone (TCL6-8) was cocultured in the presence of the 2F10 anti-Id antibody with 2F10-Fab-primed B cells. The Th clone responded with proliferation and also helped 2F10-Fab-primed B cells to produce antibodies that bind to L1210/GZL, and not to control P815 tumor cells.

Furthermore, the presentation of 2F10 anti-Id antibodies to the Th clone by primed B cells was more efficient than that by unprimed or conventional APCs. Intact 2F10 anti-Id antibodies were presented to Fc-specific Th cells by Fab (or idiotype) primed B cells more efficiently than the fragment mixture (Fab plus Fc) of 2F10 anti-Id antibodies, indicating that 2F10-Fab (or idiotype)-primed B cells captured 2F10 anti-Id antibodies through sIg receptors. The presenting B cells were sensitive to treatment with chloroquine and must have been derived from H-2-matched mice. This finding indicated that the antigen presentation by Fab-primed B cells to Fc-specific Th cells required processing and was MHC restricted (56).

Collectively, these results outlined a mechanism that may operate in anti-Id therapy of tumor-bearing animals using tumor antigen-mimicking anti-Id antibodies. A similar mechanism could be effective in tumor patients immunized with xenogeneic anti-Id antibodies operating under the intra(antigen) molecular help as outlined by Lanzavecchia and colleagues (45).

Correlation of the Idiotype-Recognizing T Cell Repertoire with Tumor Growth

Since the quality of a protective immune response against tumor growth may be determined by the quality of the anti-idiotype response, the Th cell and network in mice immunized with protective and nonprotective anti-Id antibodies was further dissected during different phases of tumor growth. This approach was based on the earlier observation that in response to antigen, total anti-idiotype activity increases before expansion and expression of the idiotype-bearing antibody response (58).

Previously, it was shown that anti-Id antibodies can induce protection against the L1210/GZL lymphoma (48–50, 59). The monoclonal antitumor antibodies 11C1 and 2B2 recognized distinct epitopes on the tumor and crossreacted with the MMTV-encoded gp52 molecule (51, 60). Here 11C1 was chosen to make anti-Id antibodies because of its high affinity for L1210/GZL and its ability to lyse tumor cells in a complement-mediated cytotoxicity assay. Also 2B2 has a lower affinity for L1210/GZL and mediates less killing of tumor cells in the presence of complement (H. Fuji, unpublished observation). Furthermore, passive transfer of the 11C1 antibody partially inhibits tumor growth. The antibodies 11C1 and 2B2 did not cross-react either idiotypically or paratypically.

During the studies with paratypic anti-Id antibodies, a differential biological response induced by the monoclonal anti-Id antibodies 2F10, 3A4, and D11 was observed (59). As discussed earlier, only presensitization of mice with 2F10 could induce protective immunity against tumor growth. The anti-Ids 3A4 and D11 were unable to induce such a protective effect. It was also demonstrated that 2F10 and 3A4 differed in their ability to induce idiotype-bearing Th cells (Th1) (59).

The main goal in this study was to analyze the Th cells that recognize the antitumor idiotypes on the Ab1s 11C1 and 2B2 in mice with tumor progression or regression. Monoclonal antitumor antibodies were used as probes for analyzing idiotype-specific Th cells (Th2). A precise knowledge of the Th2 cell response might indicate the underlying mechanism in tumor progression. A comparison of the tumor-induced regulatory network induced by protective and nonprotective anti-Id antibodies will produce information needed for more effective manipulation of the immune network, that is, the use of anti-Id antibodies as vaccines or immunoregulators.

It was demonstrated earlier that idiotype-bearing Th cells collected from mice during the early phase of tumor growth preferentially responded to nonprotective 3A4 anti-Id antibodies (59). Raychaudhuri et al (58) described the anti-Id–bearing Th cell (Th2) repertoire in Ab2-manipulated or -nonmanipulated tumor-bearing mice.

For analyzing 11C1- and 2B2-recognizing Th2 cell precursors in mice, graded numbers of lymph node T (LNT) cells were mixed with 5×10^5 splenic B cells fron TNP–KLH-immunized mice and incubated for four to five days in the presence of 11C1–TNP or 2B2–TNP. As a control antigen, UPC10–TNP was used. After incubation, the culture supernatant was collected and tested for the presence of IL-2. The results showed that, during the early phase of tumor growth, the frequency of Th cells responsive to the 2B2 idiotype was much higher than Th cell responsiveness to 11C1. T cells from DBA/2 mice (Igh^a) recognized two MAbs, 11C1 and 2B2, both generated in syngeneic DBA/2 mice (51, 60). Thus, it is unlikely that they were observing allotype-specific T cell responses in the idiotype-specific T helper assays.

Since idiotype-recognizing Th cells vary in frequency at different stages of tumor growth, it was decided to determine whether there is any difference in the frequency of Th2 cells in mice immunized with either protective or nonprotective anti-Id antibodies. The Lyt-2 + lymph node T (LNT) cells were prepared from mice immunized with protective 2F10 or nonprotective 3A4 anti-Id antibodies and incubated at different concentrations with 5×10^5

TNP-primed splenic B cells as described (59). Trinitrophenylated 11C1 or 2B2 were used as in vitro antigens and TNP-conjugated UPC10 was used as a control. After the incubation, the culture supernatant was tested for interleukin-2 (IL-2) activity. The 2F10 immunization induced more Th cells that recognized 2B2 than 11C1 that recognized Th cells. The 3A4 immunization also induced Th cells that recognized 2B2, and the frequency of 11C1-recognizing Th cells in 3A4-immunized mice was much higher than in mice immunized with 2F10. LNT cells from KLH- and UPC10-immunized mice did not recognize 2B2 or 11C1. These results indicated that tumor-related 2B2 idiotype-bearing Th cells were activated by live tumor cells and by protective or nonprotective anti-Id antibodies. The frequency of 2B2-recognizing Th cells in 3A4-immunized mice was similar to that induced by a growing tumor during the early phase of tumor growth. However, the frequency of 11C1- and 2B2-responsive Th cells in 2F10-immunized mice was less than the frequency of cells induced by 3A4 or by a growing tumor at an early growth phase.

Köhler et al (59a) also performed a study to determine whether Th2 cells induced by anti-idiotype or by growing tumor were able to cooperate with TNP-primed B cells to produce anti-TNP antibodies. It appears from the data that Lyt-2− T cells isolated from mice during the early phase of tumor growth predominantly cooperated with TNP-primed B cells to secrete antibody in the presence of 2B2-TNP as antigen, whereas comparatively lesser amounts of TNP binding antibody could be detected in the presence of 11C1-TNP. When the tumor became larger, the Lyt-2− LNT cells from these mice responded equally well with both antigens. The 2F10-induced Lyt-2− LNT cells cooperated with TNP-primed B cells only in the presence of 11C1-TNP. However, the T cells from 3A4-immunized mice cooperated only with 2B2-TNP and were therefore similar to Th cells induced by the tumor in its early growth phase. This is in contrast to the frequency data, wherein 3A4 induced more 11C1-responsive Th cells than 2F10 did. The data suggest that regulatory mechanisms are operative in mice with growing tumors and also in mice immunized with either protective or nonprotective anti-Id antibodies. The idiotype-recognizing capabilities of Th cells from mice immunized with the protective anti-Id antibody 2F10 appeared to be different from Th cells from mice immunized with nonprotective anti-Id antibodies or mice in an early phase of tumor growth.

In summary, the expression of certain idiotype-recognizing Th cells induced under different conditions correlates with the progression and regression of tumor growth. The expression of 2B2 idiotype-recognizing Th cells was found to be predominant in tumor-injected mice preimmunized with nonprotective Ab2s. However, 11C1-idiotype-recognizing Th expression was high in tumor-bearing mice preimmunized with protective Ab2s.

The differential activation of idiotype-specific Th cells by protective Ab2s could be explained differently. One possibility is that not all proliferating or IL-2-producing T cells were able to cooperate with B cells to produce Ig. Another possibility is that there was a flux in clonotypic expansion and, depending on the external stimulus, these clonotypes might be driven to a distinct differentiation step. Here, one has to assume that 2F10 can drive 11C1-recognizing T cells to a differentiation stage at which they can offer cognate help to B cells to secrete antibody. Although 2B2-recognizing T cells were activated to secrete IL-2, their differentiation state was not cooperative with B cells in the secretion of antibody. Finally, the failure to demonstrate 2F10-immunized T cells in cooperation with B cells might reflect a state of idiotype-specific suppression. The underlying mechanism by which anti-idiotypes induce differential protective immunity is unclear; however, the observed differences at the Th1 level are likely to be involved. One possibility is that 2B2 idiotype-recognizing Th cells might have induced efficient production of 3A4 and D11 idiotype-bearing regulatory T cells, which may, in turn down-regulate the effective antitumor immunity.

Altogether, these results demonstrate that different Th cells that recognize antitumor idiotypes were activated in mice at different tumor growth stages or with different patterns.

Synergistic Antitumor Effects with Combined "Internal Image" Anti-Idiotypes and Chemotherapy

In another study, Chen et al (61) explored the combination of anti-idiotypes and chemotherapy in the murine tumor system L1210/GZL for which Raychaudhuri, Köhler, and others had previously generated protective anti-Id MAbs.

At first, in this study, various protocols using a protective anti-Id in active immunization were investigated. Mice preimmunized before tumor transfer and challenged again after tumor transfer survived significantly longer. Next Chen et al (61) explored the use of soluble anti-idiotype as an immunostimulator in tumor-bearing mice. Although this treatment did not induce long-term survival, it significantly increased survival time at the optimum dose range.

Finally, stimulatory anti-Id therapy and cyclophosphamide (Cy) treatment were combined. Tumor-bearing mice were given a single dose of Cy followed by different doses of soluble anti-idiotype. An optimal effect on tumor growth and survival was achieved with 80 mg/kg of Cy and 10 μg/mouse of anti-idiotype; here 80% of the mice survived longer than 100 days. These results provide guidelines for developing clinical protocols for cancer patients using a combination therapy of anti-idiotypes and chemotherapy.

In contrast to previously described, experimentally induced anti-Id immunities (11, 62), when anti-idiotypes were given with adjuvants and often with coupled protein carriers, Chen et al (61) were able to demonstrate antitumor effects in mice with established tumors using anti-idiotypes injected intravenously and without carriers. Interestingly, the antitumor effect induced by soluble 2F10 was operative only at an optimum dose range, with higher or lower doses having no effect (61). This is reminiscent of the dose dependence in anti-Id stimulation of an nitrophenol-idiotype in mice (63, 64). Although soluble anti-idiotype alone prolonged survival but did not cure the tumor mouse, treatment with anti-idiotype combined with Cy led to long-term survival of 80% of the tumor mice. This observation of a critical dose effect of anti-idiotypes must be taken into consideration for human trials in which anti-Id antibodies are used. It should be noted, however, that Chen, Köhler, and others (61) do not yet understand the cellular mechanism that mediates the synergistic therapeutic effect of Cy and anti-idiotypes. The elucidation of this mechanism will be the subject of future studies.

Conclusion

The data and results from experimental anti-Id immunotherapy provide some new insight into the regulation of antitumor immunity in tumor hosts. Such experiences will be helpful in devising effective immunotherapy approaches using anti-id antibodies in human cancer patients.

References

1. Tilkin AF, Schaaf-Lafontaine N, Van Acker A, Boccadoro M, Urbain J: Reduced tumor growth after low-dose irradiation or immunization against blastic suppressor T cells. *Proc Natl Acad Sci USA* 1981; 78:1809.
2. Glaser M: Regulation of specific cell-mediated cytotoxic response against SV40-induced tumor associated antigens by depletion of suppressor T cells with cyclophosphamide in mice. *J Exp Med* 1979; 149:774.
3. Greene MI: The genetic and cellular basis of regulation of the immune response to tumor antigens. *Contemp Topics* 1980; 11:81.
4. Haubeck HD, Kolsch E: Regulation of immune responses against the syngeneic ADJ-PC-5 plasmacytoma in BALB-c mice. III. Induction of specific T suppressor cells to the BALB/c plasmacytoma ADJ-PC-5 during early stages of tumorigenesis. *Immunology* 1982; 47:503.
5. Howie SM, McBride WH: Tumor-specific T helper activity can be abrogated by two distinct suppressor cell mechanisms. *Eur J Immunol* 1982; 12:671.
6. McBride WH, Howie SEM: Induction of tolerance to a murine fibrosarcoma in two zones of dosage--the involvement of suppressor cells. *Br J Cancer* 1986; 53:707.
7. Köhler H, Kieber-Emmons T, Srinivasan S, et al: Revised Immune Network Concepts. 1989; *Clin Immunol Immunopathol* 1989; 52:104.
8. Weigle WO: The immune response of rabbits tolerant to bovine serum albumin to the injection of other heterologous serum albumins. *J Exp Med* 1961; 114:111.
9. Rodkey LS: Autoregulation of immune responses via idiotype network interactions. *Microbiol Rev* 1980; 44:631.
10. Urbain J, Wuilmart C, Cazenave PA: Idiotypic regulation in immune networks. *Contemp Topics Mol Immunol* 1981; 8:113.
11. Köhler H, Müller S, Bona C: Internal idiotype antigens. *Proc Soc Exp Biol Med* 1985; 178:189.
12. Sacks DL, Esser KM, Sher A: Immunization of mice against African trypanasomiasis using anti-idiotypic antibodies. *J Exp Med* 1982; 155:1108.

13. Lee VK, Harriott TG, Kuchroo VJ, Halliday WJ, Hellström I, Hellstöm KE: Monoclonal anti-idiotype antibodies related to a murine oncofetal bladder tumor antigen induce specific cell-mediated tumor immunity. *Proc Natl Acad Sci USA* 1985; 82:6286.
14. Schroer RR, Briles DE, Van Boxel JA, Davie JM: Idiotypic uniformity of cell surface immunoglobulin in chronic lymphocytic leukemia. Evidence for monoclonal proliferation. *J Exp Med* 1974; 140:1416.
15. Hough VW, Eddy RP, Hamblin TJ, Stevenson FK, Stevenson GT: Anti-idiotype sera raised against surface immunoglobulin of human neoplastic lymphocytes. *J Exp Med* 1976; 144:960.
16. Hamblin TJ, Abdul-Ahad AK, Gordon J, Stevenson FK, Stevens GT: Preliminary experience in treating lymphocytic leukaemia with antibody to immunoglobulin idiotypes on the cell surfaces. *Br J Cancer* 1980; 42:495.
17. Ritz J, Pesando JM, Notis-McConarty J, Schlossman SF: Modulation of human acute lymphoblastic leukemia antigen induced by monoclonal antibody in vitro. *J Immunol* 1980; 125:1506.
18. Hatzubai A, Maloney DG, Levy R: The use of a monoclonal anti-idiotype antibody to study the biology of a human B cell lymphoma. *J Immunol* 1981; 126:2397.
19. Miller RA, Maloney DG, Warnke R, Levy R: Treatment of B-cell lymphoma with monoclonal anti-idiotype antibody. *N Engl J Med* 1982; 306:517.
20. Sklar J, Cleary ML, Thielemans K, Gralow J, Warnke R, Levy R: Biclonal B-cell lymphoma. *N Engl J Med* 1984; 311:20.
21. Kennedy RC, Dreesman GR, Butel JS, Landford RE: Suppression of in vivo tumor formation induced by simian virus 40-transformed cells in mice receiving antiidiotypic antibodies. *J Exp Med* 1985; 161:1432.
22. Nelson KA, George E, Swenson C, Forstrom JW, Hellström KE: Immunotherapy of murine sarcomas with auto-anti-idiotypic monoclonal antibodies which bind to tumor-specific T cells. *J Immunol* 1987; 139:2110.
23. Hellström I, Hellström KE, Rollins N, Lee VK, Hudkins K, Nepom GT: Monoclonal antibodies to cell surface antigens shared by chemically induced mouse bladder carcinomas. *Cancer Res* 1985; 45:2210.
24. Nepom GT, Nelson KA, Holbeck SL, Hellström I, Hellström KE: Induction of immunity to a human tumor marker by in vivo administration of anti-idiotypic antibodies in mice. *Proc Natl Acad Sci USA* 1984; 81:2864.
25. Kahn M, Hellström I, Estin C, Hellström KE: Monoclonal anti-idiotypic antibodies related to the p97 melanoma antigen. *Cancer Res* 1989; 49:3157.
26. Greene MI, Benacerraf B: Studies on hapten specific T cell immunity and suppression. *Immunol Rev* 1980; 50:163.
27. Nisonoff A, Greene MI: Regulation through idiotypic determinants of the immune response to p-azophenylarsonate hapten in strain-A mice. *Proc 4th Int Congr Immunol* 1980; 57-80.
28. Estin CD, Stevenson US, Plowman GD, et al: Recombinant vaccinia virus vaccine against the human melanoma antigen p97 for use in immunotherapy. *Proc Natl Acad Sci USA* 1988; 85:1052.
29. Bhattacharya-Chatterjee M, Pride MW, Seon BK, Köhler H: Idiotype vaccines against human T cell acute lymphoblastic leukemia. I Generation and characterization of biologically active monoclonal anti-idiotopes. *J Immunol* 1987; 139:1354.
30. Sen L, Borella L: Clinical importance of lymphoblasts with T markers in childhood acute leukemia. *N Engl J Med* 1975; 292:828.

31. Sallan SE, Ritz J, Pesando J, et al: Cell surface antigens: prognostic implications in childhood acute lymphoblastic leukemia. *Blood* 1980; 55:395.
32. Bowman WP, Melvin SL, Aur RJA, Mauer AM: A clinical perspective on cell markers in acute lymphocytic leukemia. *Cancer Res* 1981; 41:4794.
33. Greaves MF: Analysis of the clinical and biological significance of lymphoid phenotypes in acute leukemia. *Cancer Res* 1981; 41: 4752.
34. Pullen DJ, Falletta JM, Crist WM, et al: Southwest Oncology Group experience with immunological phenotyping in acute lymphocytic leukemia of childhood. *Cancer Res* 1981; 41:4802.
35. Seon BK, Negoro S, Barcos MP, Tebbi CK, Chervinsky D, Fukukawa T: Monoclonal antibody SN2 defining a human T cell leukemia-associated cell surface glycoprotein. *J Immunol* 1984; 132:2089.
36. Seon BK, Fukukawa T, Jackson AL, et al: Human T-cell leukemia-associated cell surface glycoprotein GP37: studies with three monoclonal antibodies and a rabit antiserum. *Mol Immunol* 1986; 23:569.
37. Seon BK: Specific killing of human T-leukemia cells by immunotoxins prepared with ricin A chain and monoclonal anti-human T-cell leukemia antibodies. *Cancer Res* 1984; 44:259.
38. Hara H, Seon BK: Complete suppression of in vivo growth of human leukemia cells by specific immunotoxins: nude mouse models. *Proc Natl Acad Sci USA* 1987; 84:3390.
39. Bhattacharya-Chatterjee M, Chatterjee SK, Vasile S, Seon BK, Köhler H: Idiotype vaccines against human T cell leukemia. II. Generation and characterization of a monoclonal idiotype cascade (Ab1, Ab2, and Ab3). *J Immunol* 1988; 141(4):p1398–403.
40. Koprowski H, Herlyn D, Lubeck M, DeFreitas E, Sears HF: Human anti-idiotype antibodies in cancer patients: Is the modulation of the immune response beneficial for the patient. *Proc Natl Acad Sci USA* 1984; 81:216.
41. DeFreitas E, Suzuki H, Herlyn D, et al: Human antibody induction to the idiotypic and anti-idiotypic determinants of a monoclonal antibody against a gastrointestinal carcinoma antigen. *Curr Topics Microbiol Immunol* 1985; 119:75.
42. Koprowski H: Applications to passive and active immunotherapy in human cancer. In: *New Horizons of Tumor Immunotherapy*. The Hague, Elsevier/North Holland; in press.
43. Kusame M, Kageshita T, Tsujisake M, Perosa F, Ferrone S: Syngeneic anti-idiotypic antisera to murine anti-human high-molecular weight melanoma-associated antigen monoclonal antibodies. *Cancer Res* 1987; 47:4512.
44. Herlyn D, Wettendorff M, Schmoll E, et al. Anti-idiotype immunization of cancer patients: Modulation of the immune response *Proc Natl Acad Sci USA* 1987; 84:8055.
45. Lanzavecchia A, Abrignani S, Scheidegger D, Obrist R, Dörken B, Moldenhauer G: Antibodies as antigens. The use of mouse monoclonal antibodies to focus human T cells against selected targets. *J Exp Med* 1988; 167:345.
46. Ritz J, Schlossman SF: Utilization of monoclonal antibodies in the treatment of leukemia and lymphoma. *Blood* 1982; 59:1.
47. Shawler DL, Bartholomew RM, Smith LS, Dillman RO: Human immune response to multiple injections of murine monoclonal IgG. *J Immunol* 1985; 135:1.
48. Raychaudhuri S, Saeki Y, Fuji H, Köhler H: Tumor-specific idiotype vaccines. I. Generation and characterization of internal image tumor antigen. *J Immunol* 1986; 137:1743.

49. Raychaudhuri S, Saeki Y, Chen J-J, Iribe H, Fuji H, Köhler H: Tumor-specific idiotype vaccines. II. Analysis of the tumor-related network response inducing the tumor and by internal image antigens (Ab2b). *J Immunol* 1987; 139:271.
50. Raychaudhuri S, Saeki Y, Chen J-J, Köhler H: Tumor-specific idiotype vaccine. III. Induction of T helper cells by anti-idiotype and tumor cells. *J Immunol* 1987; 139:2096.
51. Rapp L, Fuji H: Differential antigenic expression of the DBA/2 lymphoma L1210 and its sublines: cross-reactivity with C3H mammary tumors as defined by syngeneic monoclonal antibodies. *Cancer Res* 1983; 43:2592.
52. Saeki Y, Chen J-J, Shi L, Raychaudhuri S, Köhler H: Characterization of "regulatory" idiotope-specific T cell clones to a monoclonal anti-idiotypic antibody mimicking a tumor-associated antigen (TAA). *J Immunol* 1989; 142:1046.
53. Schwartz RH: T-lymphocyte recognition of antigen in association with gene products of the major histocompatibility complex. *Ann Rev Immunol* 1985; 3:237.
54. Bottomly K, Mosier DE: Antigen-specific helper T cells required for dominant idiotype expression are not H-2 restricted. *J Exp Med* 1981; 154:411.
55. Ochi A, Worton KS, Woods G, Gravelle M, Kitagami K: A novel strategy for immunotherapy using antibody-coupled carriers to focus cytotoxic T helper cells. *Eur J Immunol* 1987; 17:1645.
56. Saeki Y, Chen J-J, Shi L, Köhler H: Idiotypic intramolecular help. Induction of tumor-specific antibodies by monoclonal anti-idiotypic antibody with the help of Fc-specific T helper clones. *J Immunol* 1989; 142:2629.
57. Lanzavecchia A: Exploiting the immune system's own strategies for immunotherapy. *Immunol Today* 1988; 9:167.
58. Raychaudhuri S, Saeki Y, Chen JJ, Fuji H, Köhler H: Tumor idiotype vaccines V. Correlation of the idiotype-recognizing T cell repertoire with tumor growth. *J Immunol*, in press.
59. Raychaudhuri S, Saeki Y, Chen JJ, Köhler H: Analysis of the idiotypic network in tumor immunity. *J Immunol* 1987; 139:3902.
59a. Köhler H, Raychaudhuri S, Chen J-J, Saeki Y: Idiotypic responses induced by tumor: An autocrine network. *Intern Rev Immunol* 1989, 4:311.
60. Fuji H, Iribe H: Clonal variation in tumorigenicity of L1210 lymphoma cells: nontumorigenic variants with an enhanced expression of tumor-associated antigen and Ia antigens. *Cancer Res* 1986; 46:5541.
61. Chen J-J, Saeki Y, Shi L, Köhler H: Tumor idiotype vaccines VI: Synergistic antitumor effects with combined "internal image" anti-idiotypes and chemotherapy. *J Immunol*, 1989; 143:1053.
62. Kennedy RC, Dreesman GR, Köhler H: Vaccines utilizing internal images anti-idiotypic antibodies that mimics antigens of infectious organisms. *Biotechniques* 1985; 3:4040.
63. Kelsoe G, Roth M, Rajewsky K: Control of idiotype expression by monoclonal anti-idiotope and idiotope-bearing antibody. *Eur J Immunol* 1981; 11:418.
64. Rajewsky K, Takemori T: Genetics, expression, and function of idiotypes. *Ann Rev Immunol* 1983; 1:569.

CHAPTER 7

Anti-Idiotypic Antibodies as Potential Viral Vaccines

Ronald Q. Warren and Ronald C. Kennedy

Introduction

Although conventional vaccines have been demonstrated to be highly effective in stimulating protective immunity against a wide range of microorganisms, there are still instances in which suitable vaccines are lacking. For example, the inability to culture large quantities of numerous protozoal pathogens in vitro has prevented the development of conventional vaccines for these organisms. Other obstacles to the development of conventional vaccines include the poor humoral response of infants under the age of 2 years to certain polysaccharide antigens present on encapsulated bacteria (i.e., *Neisseria meningitidis, Vibrio cholera,* and *Hemophilus influenzae*) (1, 2). Also, the recent spread of the human immunodeficiency virus type 1 (HIV-1) and its possible origin and evolution from viruses isolated from nonhuman primate species has pointed up the necessity of developing a vaccine that is incapable of reverting to a virulent form. These obstacles have led to the development of a new generation of vaccines. Recombinant proteins, synthetic peptides, and anti-idiotypic antibodies represent relatively new technologies that are currently being examined for their potential use in stimulating immunity against various microorganisms (reviewed in 3 and 4). This report will review studies from this laboratory that examine the potential use of anti-Idiotype to generate immunity to HIV-1 and hepatitis B virus (HBV).

The idiotypic approach to vaccine development is based on the presence of immunogenic regions, termed *idiotypes,* which are located on the antigen receptors of both B and T lymphocytes. Multiple idiotypes can be located within a given variable (V) region. Since T cell receptors and immunoglobulin molecules represent unique arrangements of V, D, and J region genes, the corresponding Id can serve as a marker for specific lymphocyte and antibody populations. Antibodies reactive to these idiotypic regions are termed *anti-Idiotypes.* Jerne first described how the immune system can potentially be regulated through the interactions of Id + lymphocytes with cells that possess complementary anti-idiotype receptors (5). These interactions can have either stimulatory or inhibitory effects on lymphocyte activity (reviewed in 6 and 7).

During normal periods, these positive and negative signals are thought to be in a state of equilibrium, with the lymphocyte remaining in a resting state. When challenged with foreign antigen, an increase in positive signals results in the activation and expansion of particular Id + lymphocyte clones. Thus, lymphocytes may respond not only to foreign antigen but also to idiotopes present on autologous lymphocytes. The immune system can therefore be represented as a network of lymphocyte interactions regulated through the recognition of antigen, idiotypes, and anti-idiotypes.

Several types of anti-idiotypes have been characterized based on specificity and the ability to mimic the three-dimensional conformation of native antigen. According to Jerne, two kinds of anti-idiotypes can be distinguished: Ab2α and Ab2β (8). The Ab2α binding to its Ab1 may or may not be inhibited by the antigen used to induce the Ab1. The Ab2β, or internal image anti-idiotype, has the capacity to mimic the antigen used to generate the Ab1 serologically and can substitute for the native antigen. An alternative classification of anti-idiotype was proposed by Bona and Kohler (9). According to their classification, the Ab2α subclass is divided into anti-idiotypes that do not recognize the antigen-combing site (Ab2α) and anti-idiotypes that react with idiotype near or within the antigen-combining site (Ab2γ). An Ab2β can mimic the three-dimensional structure of antigen and can inhibit antigen binding to Ab1. Because of the structural mimicry of Ab2β to antigen, antisera raised to native antigen in a variety of species can cross-react with Ab2β. Although instances of amino acid sequence homology between native antigen and Ab2β immunoglobulin variable regions have been reported, it is more likely that the antigenic mimicry observed represents similarities in protein folding and amino acid charge patterns rather than sequence homology (10). The Ab2γ also reacts to idiotopes located at or near the antigen-combining site of Ab1. Although Ab2γ can inhibit the binding of antigen to Ab1, it is not recognized by antisera produced in other species against the native antigen. Recent studies based on in vivo modulation experiments suggest that Ab2γ and Ab2α may not represent distinct subclasses of anti-idiotypes but rather serological variants of the same subclass (Anderson SA, Kennedy RC, 1989. Unpublished observation). A fourth type of anti-idiotype, termed Ab$_{2\varepsilon}$ or *epibody*, has also been described and has been characterized, based on its ability to bind the idiotype and the epitope on the antigen. These types of anti-idiotypes have been described on rheumatoid factors, which recognize idiotopes located on autologous IgG molecules (11).

Immunization of animals with anti-idiotype has been shown to lead to either the activation or suppression of Id + lymphocyte clones. Important factors that influence these immunomodulatory effects include dosage, method of presentation of Ab2, and the stage of development of the animal's immune system. For example, injection of newborn animals with anti-idiotype has been reported to decrease circulating levels of Id + lymphocytes (12, 13). Injection of Ab2 has also been reported to stimulate cellular immunity in vivo (14–17). Investigators have reported the induction of antigen-specific cytotoxic T cells following immunization of mice with anti-idiotype (18). There is evidence that suggests that anti-idiotype immunization can activate both B and T lymphocytes, possibly through a cross-reactive Idiotype on their respective antigen receptors.

Anti-idiotype–based vaccines have been reported to stimulate immunity against a wide range of viral antigens in a number of animal models. Injections of anti-idiotypes have led to the induction of immune responses specific for a number of viral antigens (see Table 7.1) (14, 19–36). These types of vaccines could potentially circumvent several of the obstacles that appear to plague conventional vaccines. For example, internal image anti-idiotype can mimic

Table 7.1. Summary of anti-idiotype used as a potential vaccine in animal models.

Virus	Comment	Reference(s)
Reovirus		
	Monoclonal Ab2β, stimulated DTH and CTL response	14, 19
HIV		
	Polyclonal Ab2, stimulated anti-gp160 response, activated silent clones	20
	Anti-Leu3a immune mice sera, recognized gp160 and inhibited syncytia formation	21
Hepatitis B virus		
	Polyclonal Ab2, protected chimpanzees against challenge with hepatitis B virus	22
	Polyclonal Ab2, induced antiviral antibodies	23
	Polyclonal Ab2α, induced antiviral antibodies, activated silent clones	24
	Monoclonal Ab2β, induced antiviral antibodies	25, 26
Cytomegalovirus		
	Monoclonal Ab2β, induced CMV-neutralizing antibodies	27
Sendai virus		
	Monoclonal Ab2, induced CTLs, protected against lethal challenge	28, 29
Rabies virus		
	Polyclonal Ab2, induced viral-neutralizing antibodies	30
Polio virus		
	Monoclonal Ab2, induced polio virus-neutralizing antibodies	31
Herpes simplex virus		
	Polyclonal Ab2, induced DTH	32
	Polyclonal Ab2, induced antiviral antibodies	33
Influenza virus		
	Polyclonal Ab2, induced antiviral antibodies	34
Newcastle disease virus		
	Polyclonal Ab2, induced antiviral antibodies	35
Coxsackievirus		
	Polyclonal Ab2, induced cell-mediated responses	36

epitopes on viruses that are difficult to grow in vitro. Certain classes of anti-idiotypes have been shown to be immunogenic across species barriers and should be no more genetically restricted than the corresponding native antigen (19, 22, 29). Anti-idiotype–based vaccines would also eliminate the risk of introducing infectious material (including retroviral nucleic acids) into individuals during immunization, since no living infectious materials are involved in anti-idiotype preparations. The anti-idiotype vaccine approach would also bypass the need to culture potentially dangerous microorganisms. Another advantage would occur when immunity to a single epitope on an infectious agent is adequate to induce protection. Some infectious agents may contain antigenic determinants in common with host tissue, which would lead to the development of a cross-reactive autoimmune response. An anti-idiotype could theoretically induce immunity to a single epitope and bypass a potential autoimmune response. The ability of anti-idiotype to mimic a nonproteinaceous antigen offers another potential advantage. Studies that suggest that this potential exists are based on the capsular polysaccharides of pathogenic bacteria (reviewed in 4). In some instances, the anti-idiotype can mimic the capsular polysaccharide, such that individuals who normally do not develop an immune response to these antigens until several years after birth may respond to the protein-based anti-idiotype and produce an antipolysaccharide response. This raises the possibility that anti-idiotype–based vaccines may have applications in certain neonatal infections. Also, anti–idiotype may modulate the immune response to selected viral epitopes. One can specify an epitope from which the most desirable immune response would result and possibly enhance or prime the production of antibodies specific for this viral epitope. Another possible advantage of anti-idiotype–based vaccines occurs when the organism exhibits a high degree of genomic diversity. Here anti-idiotype can be used to induce an immune response to conformational determinants, such as sites of interactions with cell surface receptors. The potential for anti-idiotypes to mimic viral epitopes, which are important for the recognition of the CD4 molecule, provides the basis for the potential use of anti-idiotypes to inhibit HIV infection. A final advantage would be gained during induction of immunity against toxins in which the nature of these compounds precludes their use for immunization (37,38). The potential advantages of anti-idiotype–based vaccines strategies are summarized in Table 7.2.

The ability of anti-idiotype to activate normally silent lymphocyte clones indicates that this procedure may potentially direct an immune response against weakly immunogenic, neutralizing epitopes (20, 24, 39–43). Silent clones represent groups of lymphocytes that are not normally activated when an individual is immunized with foreign antigen. The potential to activate silent clones may provide a means to induce protective immunity in an individual who is normally unresponsive to a particular antigen. One potential application for this type of immunization involves generating immunity to the HIV. Individuals infected with HIV typically mount a strong

Table 7.2. Advantages of anti-idiotype based vaccine strategies.

1. Eliminate risk of introducing infectious material along with vaccine
2. Eliminate need to culture microorganisms that are difficult and/or hazardous to grow in vitro
3. Internal image anti-idiotype may mimic epitope on a nonproteinaceous, weakly immunogenic antigen (i.e., polysaccharides, carbohydrates, glycolipids)
4. Internal image anti-idiotype may mimic cellular receptor for virus, thereby blocking viral infectivity (i.e., mimicry of CD4 molecule)
5. Internal image anti-idiotype may mimic epitopes on toxins, allowing immunity to be generated without exposure to lethal agents
6. Internal image anti-idiotype may direct an immune response toward a single, neutralizing epitope, which may not be immunodominant on the native organism

---

humoral immune response to viral antigens. Although these antibodies may exhibit neutralizing activity in vitro, they typically do not halt the spread of the virus or disease progression in vivo. Activation of protective lymphocyte clones via idiotype–anti-Id interactions could have the potential to lead to a more effective immune response against HIV and might represent a new type of therapeutic strategy.

Use of Anti-Idiotype to Modulate HIV-1 Immunity

Acquired immunodeficiency syndrome (AIDS) was first recognized as a new disease affecting the immune system in 1981 (44–46). The etiological agent of this disease is a type C human retrovirus, which has been collectively referred to as HIV. The complete genome of several isolates of HIV have been cloned and sequenced (47–50). HIV is composed of single-stranded RNA encased by a nucleocapsid material. The nucleocapsid of the virus is surrounded by a lipid bilayer envelope, which is composed of glycoproteins gp120 and gp41. These glycoproteins are cleaved from a precursor, gp160, which is encoded by the HIV genome. The cellular receptor for HIV-1 is the CD4 molecule.

The CD4 molecule is a T cell surface glycoprotein that has been implicated in mediating lymphocyte–target cell interactions. It is thought that the CD4 molecule is recognized by major histocompatibility complex (MHC) class II antigens (51–58). The CD4 molecule is preferentially expressed on the surface of the T helper/inducer subset of lymphocytes (59) and may bind monomorphic epitopes of the MHC in order to enhance the avidity of T cell antigen-specific recognition (60, 61). The CD4 molecule is believed to be involved in cell fusion, or syncytia formation, which allows HIV-infected cells to fuse with uninfected cells expressing CD4 molecules on their cell membranes.

The biochemical and molecular characterization of the human CD4 and murine L3T4 molecules suggest that both are members of the immunoglobulin supergene family (62, 63). The nucleotide and predicted amino acid sequences of CD4 indicate that the molecule possesses a 23 amino acid leader sequence, which is followed by a V-like region. This V region, together with the amino terminal portion of CD4, exhibits a 32% homology with the immunoglobulin light chain and contains one intrachain disulfide bond (cysteine residues at positions 16 and 84). Studies have shown that the CD4 V_1 domain serves as a receptor for HIV gp120 (64). The CD4 molecule also appears to exhibit a joining (J)-like region, which displays sequence homology with both the immunoglobulin J and T cell receptor (TCR) regions. A third extracellular domain appears to be unrelated to any known protein and exhibits two potential glycosylation sites containing the sequence Asn–Leu–Thr, which may allow for potential N-linked glycosylation. The CD4 molecule also exhibits a transmembrane portion, which contains approximately 50% homology with the class II MHC B-chain and a highly charged cytoplasmic region of approximately 40 amino acids. It is not unreasonable to speculate that, based on structural similarities of CD4 to immunoglobulin, the CD4 structure may be involved in cell–cell interactions. The extracellular sequence of the CD4 molecule includes six cysteine residues at amino acid positions 16, 84, 130, 159, 302, and 345. These cysteines may represent potential sites for intrachain disulfide bond formation. Such a loop structure is reminiscent of immunoglobulin-like domains. The molecular, biochemical, and functional characteristics of the murine L3T4 hares significant homology with the human CD4. The L3T4 molecule contains three disulfide bridges between adjacent pairs of cysteine residues (65). The possibility exists that similar linkages may occur in the human CD4 molecule. Thus, the CD4 molecule may contain four immunoglobulin-like domains (66,67).

An important consideration in the design of vaccine strategies for HIV concerns the high degree of genetic polymorphism between isolates, especially within gp120 (68). Hypervariability within envelope gene products has frequently been observed in lentiviruses (69). Sequencing of HIV-1 gp160 demonstrates that the majority of isolates from the United States differ by approximately 20% in amino acid composition (70). Because of this variability, neutralizing antibodies generated against one isolate may not be reactive with other strains (68,71,72). Indeed, studies have indicated that humoral immune responses induced to purified HIV envelope glycoproteins in experimental animals exhibit predominantly type-specific neutralizing activity in vitro (73). This antibody fails to neutralize genetically divergent isolates of HIV. Recent reports have also indicated that HIV-1 variation in vivo is rapid and that a relatively large number of related but distinguishable genetic variants evolve in parallel and can coexist during chronic infection (74,75). An effective vaccine should contain all potential group-specific neutralizing epitopes but not the potentially divergent hypervariable regions of the virus.

Another problem associated with developing a vaccine for HIV is that the infection may spread by both free virus and infected cells. A vaccine candidate should therefore induce both cell-mediated immunity and antibodies capable of mediating antibody-dependent, cell-mediated cytotoxicity (ADCC). Both types of immunity may be critical to the elimination of HIV-infected cells.

Two other concerns are noteworthy and should be considered in the design of HIV vaccine strategies. First, HIV may latently infect cells and thereby avoid the immune system by becoming dormant within cells. Second, an immune response may be produced that enhances viral infectivity by promoting the uptake of virus–antibody complexes into Fc receptor-bearing cells, such as monocytes and macrophages. Earlier studies have shown that monocytes and macrophages are permissive for HIV infection (76). Viral-enhancing antibodies have been previously described during infection with Dengue virus (77). Recently, in vitro antibody-dependent enhancement of HIV-1 infection has been reported in sera from some seropositive individuals (78,79).

Many laboratories are currently pursuing the development of a safe, effective vaccine for AIDS. A wide range of approaches to an HIV vaccine are being considered, including the use of recombinant HIV proteins, synthetic peptides, inactivated virus, infusion of soluble CD4, monoclonal antibodies to CD4, and anti-idiotypes (reviewed in 80–82). A successful AIDS vaccine will have to overcome the obstacles mentioned above (summarized in Table 7.3).

Our laboratories have been actively involved in analyzing two forms of idiotype-based HIV vaccines. The first centers on immunizing animals with monoclonal anti-CD4 antibodies. A portion of the anti-idiotype generated against anti-CD4 antibodies (Ab1) may represent the internal image of the CD4 molecule. Under these conditions, the continuous production in vivo of internal image anti-idiotype that serologically mimics CD4 could potentially lead to the saturation of CD4-binding sites on HIV gp120. Other investigators have demonstrated that soluble CD4 protein can inhibit HIV binding to CD4 + cells in vitro (83–87). This type of approach might prevent infection of CD4 + cells by HIV. Since both HIV-1 and HIV-2 appear to utilize the CD4 molecule for cell binding, a conserved conformation may be responsible for a gp120–CD4 interaction. A CD4 internal image anti-idiotype that interferes with this recognition site could potentially neutralize a wide spectrum of HIV isolates. By stimulating the continuous synthesis of high titers of CD4 internal image anti-idiotype in vivo, the need for multiple infusions of soluble CD4 might be bypassed.

A panel of approximately 25 monoclonal anti-CD4 antibodies were initially screened for their ability to competitively inhibit each other's binding to a CD4 positive T-cell line. Multiple epitopes on the CD4 molecule were identified using this procedure. Each monoclonal anti-CD4 was grouped according to epitope reactivity. These antibodies were then tested for their ability to inhibit

Table 7.3. Difficulties in developing an effective vaccine against AIDS.

1. Genetic variability among different HIV isolates
2. Individual can be infected with multiple isolates
3. Induction of primarily type-specific neutralizing antibodies
4. Transmission may be cell free or cell associated
5. Latency of the virus within infected cells
6. Potential for viral enhancement as a result of the immune response against HIV

HIV-1 binding to susceptible T cell lines in vitro. These results suggested that several CD4 epitopes may be involved in HIV recognition, since anti-CD4 with differing CD4 epitope specificities were found to inhibit binding of HIV-1 (88).

Our laboratories generated a mouse monoclonal anti-idiotype against anti-Leu3a (a murine monoclonal anti-CD4 preparation that blocks the gp120–CD4 interaction). This anti-idiotype demonstrated the following characteristics: (a) it bound anti-Leu3a but did not bind a panel of irrelevant murine monoclonal antibodies; (b) it partially inhibited staining of CD4-positive T cells by anti-Leu3a; (c) it reacted with HIV antigens in a commercial enzyme-linked immunosorbent assay; (d) it reacted by immunofluorescence with HIV-infected human T cells but not with uninfected T cells; (e) it bound to a molecule of 120 kDa in Western blot analysis, which utilized HIV-infected cell lysates; (f) the reaction of anti-Leu3a to the anti-idiotype was inhibited by several murine monoclonal anti-CD4 antibodies that inhibited HIV infectivity in vitro; and (g) it partially neutralized HIV infectivity of T cell lines in vitro (89). These results suggest that this anti-idiotype reacted with an idiotype determinant on anti-Leu3a and serologically mimicked part of the CD4 molecule that represents the viral receptor for HIV-1.

Our laboratory has also produced a polyclonal anti-idiotype response to anti-Leu3a in BALB/c mice. This polyclonal anti-idiotype exhibited in vitro neutralizing activity against four divergent HIV-1 isolates (HTLV-IIIB, SF-2, MN, and RF) along with an HIV-2 isolate (HIV-2_{ROD}). The anti-idiotype recognized anti-Leu3a but failed to bind an anti-CD4 preparation (OKT4), which does not inhibit HIV–gp120 binding to the CD4 molecule. In addition, the anti-idiotype bound gp160 in a solid phase immunoassay (21). Together, these results suggest that in some instances anti-CD4 antibodies may be useful in inducing an anti-Id response in mice with the capacity to bind HIV at the CD4 recognition site.

Baboons immunized with several mouse monoclonal anti-CD4 antibodies, either individually or as a cocktail preparation, also produced anti-idiotype responses. Depending on the monoclonal anti-CD4 preparation utilized for immunization, the baboon anti-idiotype response detected either a private or a cross-reactive idiotype on a series of 12 monoclonal anti-CD4 antibodies produced by the Becton Dickinson Monoclonal Center (Mountain View, CA). Baboons immunized with the cocktail anti-CD4 preparation had a broadening of the cross-reactive idiotype response. The anti-idiotype response in baboons appeared to recognize an antigen-combining site–related idiotype based on inhibition analysis. Western blot analysis indicated that gp120-reactive antibodies were transiently present following immunization with the monoclonal anti-CD4 in some of the baboons. The anti-HIV-1–gp120 response also appeared to recognize a cross-reactive epitope expressed on simian immunodeficiency virus (SIV) gp120. This recognition of a shared epitope on HIV-1 and SIV gp120 may have resulted from the presence of an internal image population of anti-idiotype that serologically mimics CD4.

Peripheral blood lymphocytes (PBL) obtained from the anti-CD4 immunized baboons were examined for their surface phenotypes and in vitro proliferative response to mitogens. Their values were comparable to those of normal baboon PBL, which suggests that anti-CD4 immunization did not cause lymphocyte depletion or energy in these baboons.

In addition, we also immunized rhesus monkeys with monoclonal anti-CD4 preparations. The anti-Id response obtained in rhesus monkeys was serologically similar to that induced in baboons. An anti-SIV–gp120 response was observed in some of the rhesus monkeys following multiple intramuscular immunizations with the anti-CD4 preparations. Overall, these studies suggest that murine monoclonal anti-CD4 could be used to induce anti-HIV-1 gp120-specific responses that cross-react with epitopes on SIV gp120. However, further studies are required to assess the application of monoclonal anti-CD4 preparations as idiotype-based vaccines and/or therapeutics for controlling HIV infection and AIDS.

A second type of anti-idiotype–based HIV vaccine being investigated by our laboratory centers on the immunization of animals with an anti-idiotype that may mimic epitopes present on the envelope of HIV. The resulting Ab3 may potentially react with both the anti-idiotype and HIV antigens. Chimpanzees were initially immunized with a synthetic peptide corresponding to amino acid sequences 735 to 752 of gp160, which is contained within the transmembrane glycoprotein gp41. The chimpanzees produced antibodies that were reactive to both HIV gp41 and gp160. Antiserum was then affinity purified by passage over an immunosorbent column containing the gp41 synthetic peptide. The anti-peptide antisera were used to immunize rabbits in order to generate anti-idiotype. The rabbit anti-idiotype reacted to an idiotype located near the antigen-combining site of the chimpanzee Ab1. This anti-idiotype could partially inhibit the chimpanzee Ab1 from binding to the HIV peptide (20). Mice immunized with rabbit anti-idiotype produced antibodies that recognized both the HIV synthetic peptide and the recombinant gp160. Interestingly, the idiotypes present on the mouse anti-HIV antibodies are not normally expressed in mice immunized with either HIV gp160 or the synthetic peptide. These experiments demonstrated that normally silent anti-HIV lymphocytes clones could be activated by immunization with anti-idiotypes. Since anti-idiotypes have been shown to activate T lymphocytes, both humoral and cellular immunity to HIV could potentially be generated by this procedure.

We have also generated and characterized a monoclonal anti-idiotype against the chimpanzee anti-gp41–Ab1 preparation specific for synthetic peptide 735-752. This monoclonal anti-idiotype appears to represent a noninternal image subclass of anti-idiotypes. Characteristics of the monoclonal anti-Id preparation include: (a) the ability to recognize the homologous chimpanzee anti-gp41 preparation along with a second heterologous chimpanzee anti-gp41; (b) the failure to recognize an interspecies idiotype expressed on rabbit and murine monoclonal antibodies generated against a gp160 synthetic peptide 735-752; and (c) the inability to inhibit the chimpanzee Ab1 binding to synthetic peptide 735-752 (Zhou and Kennedy, submitted for publication). BALB/c mice and rabbits were immunized with this monoclonal anti-idiotype preparation. Immunized mice produced an anti-Id (Ab3) response that recognized peptide 735-752 and a recombinant gp160 preparation by ELISA. Sera from these mice inhibited the idiotype–anti-Id reaction, which indicated that an Id + Ab3 was induced. The murine Ab3 also detected gp41 by Western blot analysis. Again, the murine Ab3 response expressed an idiotype that was not normally expressed in mice immunized with either gp160 or the synthetic peptide 735-752. Thus, silent idiotype clones appeared to be activated in BALB/c mice immunized with this anti-Id preparation. These data suggest the potential use of noninternal image anti-idiotype to modulate a particular anti-HIV immune response. When rabbits were immunized with the monoclonal anti-idiotype preparation, they produced an Ab3 response that failed to bind either peptide 735-752 or HIV gp160 but expressed an idiotype reactive with the anti-idiotype preparation. Thus, a noninternal image anti-idiotype can modulate antigen- and/or idiotype specific responses that are not normally induced by immunization with the nominal antigen. The ability to alter the natural immune response to HIV through idiotype–anti-Id interactions opens up new possibilities in the design of AIDS therapies.

Use of Anti-Idiotype to Induce Immunity to Hepatitis B Virus

Hepatitis B virus infects millions of people annually and represents a major worldwide health problem, especially in third world countries. Infection with HBV is often associated with the development of chronic liver disease and primary hepatocellular carcinoma (90). Fortunately, the majority of infected individuals recover completely, with the virus being eliminated and the hepatic injury resolved. However, an infectious chronic carrier state occurs in 5 to 10% of HBV infections. In most patients who are chronic carriers, biochemical and histological abnormalities are minimal to absent, but in others, hepatocellular injury may progress to cirrhosis and liver carcinoma. It has been demonstrated that antibodies against hepatitis B surface antigen (HBsAg) are protective against infection, whereas antibody to the core antigen is not protective (91). Serologically, HBsAg contains within its structure a group-specific, cross-reacting determinant(s), termed *a*, and two sets of allelic subtype determinants, termed *d* or *y* and *w* or *r*. Combinations of the *a* determinant with the various allelic subtype determinants result in four major serotypes associated with HBsAg: *adw*, *ayw*, *adr*, and *ayr* (92–94). The group-specific *a* determinant(s) has been shown to induce protective antibodies against HBV; thus, immunization with one HBsAg serotype confers protection against other HBV serotypes (95).

We have also examined the role of idiotype network interactions during the humoral immune response to HBsAg. Sera from two humans who were positive for HBsAg-reactive antibodies were passed over an affinity column containing purified HBsAg. Antibodies reactive with HBsAg (anti-HBs) were eluted and subsequently used to immunize rabbits to produce anti-Id antibodies. The binding of this rabbit anti-Id to the human anti-HBs (Ab1) preparation could be inhibited by purified HBsAg, which suggests that the anti-Id was reactive with an idiotype located at or near the Ab1 antigen-combining site (96-98).

The polyclonal anti-idiotypes were found to react with a high percentage of sera from humans infected with HBV; they were not, however, reactive with normal sera. The presence of a cross-reactive idiotype (CRI) on anti-HBs antibodies in humans suggested that a common V-region immunoglobulin gene was being utilized to produce the anti-HBs response. This anti-idiotype also recognized a common CRI present on anti-HBs from several other species (99). These data suggest that the anti-idiotype exhibited Ab2β-like activity and could possibly be substituted for HBsAg.

We then examined whether this anti-Id preparation could be used to induce formation of anti-HBs Ab3 in mice. Injection of BALB/c mice with this anti-idiotype led to an increased number of anti-HBs plaque-forming cells in spleens from these animals (100). Further analysis showed that the Ab3 were reactive with the HBsAg group-specific *a* determinant (23). These experiments, along with other published studies, demonstrated that HBsAg-reactive antibodies could be generated in mice and rabbits following immunization with anti-idiotype (101-103).

This anti-idiotype was also tested in chimpanzees for its ability to induce anti-HBs. Two chimpanzees received four injections of anti-idiotype over an 11-week period. Anti-HBs were detected in both animals at week 12 by commercially available immunoassays. Both chimpanzees were then challenged with infectious HBV at week 24 and were followed clinically until week 70. These animals were shown to be protected from HBV infection by serological and biochemical analyses. Two control chimpanzees, one untreated and the other receiving normal rabbit IgG, both developed serological and biochemical signs of HBV infection by 5 weeks after challenge (22). These studies are summarized in Table 7.4.

An important point concerning any type of potential vaccine involves its ability to stimulate the persistant production of high titers of protective antibodies. In addition to an initial immune response to the selected antigen, memory cells must be activated so that long-term immunity can be achieved. The anti-Id–immunized chimpanzees developed high titers of anti-HBs that are still present three years after HBV challenge. These results demonstrate that an anti-Id antibody preparation can be used as a potential vaccine to protect chimpanzees against HBV and to stimulate lasting immunity.

A rabbit polyclonal anti-idiotype was also generated against a murine monoclonal antibody specific for the HBsAg *a* determinant (24). This anti-Id

Table 7.4. Characteristics of a polyclonal anti-Id preparation that exhibits serological mimicry of HBsAg.

1. Detects a common idiotype on human anti-HBs from HBV-infected and HBsAg-immunized individuals (96)
2. Recognizes an antibody-combining site idiotype associated with an anti-HBs response specific for the *a* determinant of HBsAg (97)
3. Detects an interspecies idiotype on anti-HBs produced in mammals by HBsAg immunization (98)
4. Fine specificity of an idiotype paratope is centered around amino acid sequences 117 to 137 from the S region of HBsAg (98)
5. Primes the anti-HBs response to HBsAg or synthetic peptides in mice (23, 100–103)
6. Induces an anti-HBs response in mice, rabbits, and chimpanzees without subsequent HBsAg immunization (102, 103)
7. Protects chimpanzees from an infectious HBV challenge (22)
8. The specificity of the anti-idiotype induces an anti-HBs (Ab3) response that serologically resembles that of the human anti-HBs Ab1 (22)

preparation was identified as an Ab2α, since it was idiotype specific but would not inhibit HBsAg binding to the monoclonal anti-HBs Ab1. Further analysis indicated that this idiotype was not expressed on a majority of mouse monoclonal anti-HBs preparations specific for the *a* determinant.

Immunization of BALB/c mice with the Ab2α preparation led to the production of anti-HBs antibodies. Sera from these mice were found to express the idiotype present on the original murine monoclonal anti-HBs (Ab1). Absorption of these sera with HBsAg was shown to remove Id+ antibodies, confirming that this idiotype was present on the anti-HBs. These experiments demonstrated that an Ab2α could be used to induce an anti-HBs response in BALB/c mice. Therefore, an anti-HBs immune response could be induced in mice by immunization with either an internal image (Ab2β) or noninternal image (Ab2α) polyclonal anti-idiotype.

Conclusions

The use of anti-idiotypes as immunodulators has several potential clinical applications. First, anti-idiotype can be used to prime the immune system prior to exposure to antigen. Many investigators have demonstrated that the injection of an anti-idiotype can lead to the activation of complementary Id+ lymphocytes in vivo (for review see 104). The immune response to a particular antigen could thereby be selectively dominated by particular Id+ antibodies. In this manner, one could select an idiotype found on protective antibodies and produce an anti-idiotype that was reactive with the specific Ab1. This anti-idiotype could potentially be used to direct the humoral immune response toward a neutralizing viral or bacterial epitope. Potential clinical applications for this approach to vaccination include neonates who typically respond poorly to polysaccharide antigen present on such organisms as *Hemophilus influenzae* (105, 106).

In addition, studies with HIV-1-infected individuals have shown that, despite the presence of high titers of anti-HIV antibodies, the disease still progresses. If protective antibodies to HIV-1 can be identified in humans, these lymphocytes may be activated in vivo through idiotype–anti-Id interactions. Thus, idiotype-based strategies may have potential application to produce vaccines for individuals who have not been infected and immunotherapies for those individuals who are chronically infected. As we have shown, several experimental examples indicate the potential use of anti-idiotype as an alternative to vaccination. Potential problems do exist however, and may arise from the injection of a heterologous antibody preparation into humans. This is especially true if multiple injections are required, since an immune response to the heterologous antibody may decrease the effectiveness of the anti-Id vaccination or produce deleterious side effects that result from allergic reactions or immune complex formation. It may be possible to bypass this problem if a single injection is adequate to prime or enhance the immune response. In addition, methods are now becoming available to combine the V region of a murine antibody to the constant region of a human antibody. Such chimeric antibodies may be less immunogenic to humans while maintaining their idiotype specificity. The application of anti-idiotype as an alternative vaccine for infectious diseases is still in its infancy–an infancy with great potential.

Acknowledgments. This work was supported by grants AI22380, AI22307, and AI26492 from the National Institutes of Health.

References

1. Pincus DJ, Morrison D, Andrews C, Lawrence E, Sell SH, Wright PF: Age-related response to two *Haemophilus influenzae* type b vaccines. *J Pediatr* 1982; 100:197–201.
2. Parke JC Jr, Schneerson R, Robbins JB, Schlesselman JJ: Interim report of a controlled field trial immunization with capsular polysaccharides of *Haemophilus influenzae* type b and group C *Neisseria meningitidis* in Mecklenburg County, North Carolina (March 1974–March 1976). *J Infect Dis* 1977; (suppl) 136:s51–s56.
3. Arnon R: Peptides as immunogens: Prospects for synthetic vaccines. *Curr Top Microbiol Immunol* 1986; 130:1–12.
4. Hiernaux JR: Idiotypic vaccines and infectious diseases. *Infect Immun* 1988; 56:1407–1413.
5. Jerne NK: Toward a network theory of the immune system. *Ann Immunol (Paris)* 1974; 125C:373–389.
6. Bona CA, Pernis B: Idiotypic networks. In: Paul WE, ed: *Fundamental Immunology.* New York: Raven Press; 1984; 577–592.

7. Bona CA ed: *Biological Applications of Anti-Idiotypes, II.* Boca Raton, Fla: CRC Press; 1988.
8. Jerne NK, Roland J, Cazenave P-A: Recurrent idiotopes and internal images. *EMBO J* 1982; 1(2):243–247.
9. Bona CA, Kohler H: Anti-idiotypic antibodies and internal images. In: Venter JC, Fraser CM, Linstrom J, eds: *Monoclonal and Anti-Idiotypic Antibodies: Probes for Receptor Structure and Function.* Receptor Biochemistry and Methodology Series, vol. 4. New York; Alan R. Liss; 1984: p 141–149.
10. Bruck C, Co MS, Slaoui M, et al: Nucleic acid sequence of an internal image-bearing monoclonal anti-idiotype and its comparison to the sequence of the external antigen. *Proc Natl Acad Sci USA* 1986; 83:6578–6582.
11. Bona CA, Finley S, Waters S, Kunkel HG: Anti-immunoglobulin antibodies. III. Properties of sequential anti-idiotypic antibodies to heterologous anti-γ globulins. Detection of reactivity of anti-idiotype antibodies with epitopes of Fc fragments (homobodies) and epitopes and idiotopes (epibodies). *J Exp Med* 1982; 156:986–999.
12. Augustin A, Cosenza A: Expression of new idiotypes following neonatal idiotypic suppression of a dominant clone. *Eur J Immunol* 1976; 6:497–501.
13. Bona C, Stein KE, Lieberman R, Paul WE: Direct and indirect suppression induced by anti-idiotype antibody in the insulin-bacterial levan antigenic system. *Mol Immunol* 1979; 16:1093–1101.
14. Sharpe AH, Gaulton GN, McDade KK, Fields BN, Greene MI: Syngeneic monoclonal antiidiotype can induce cellular immunity to reovirus. *J Exp Med* 1984; 160:1195–1205.
15. Nepom GT, Nelson KA, Holbeck SL, Hellström I, Hellström KE: Induction of immunity to a human tumor marker by *in vivo* administration of anti-idiotypic antibodies in mice. *Proc Natl Acad Sci USA* 1984; 81:2864–2867.
16. Forstrom JW, Nelson KA, Nepom GT, Hellström I, Hellström KE: Immunization to a syngeneic sarcoma by a monoclonal auto-anti-idiotypic antibody. *Nature* 1983; 303:627–629.
17. Lee VK, Harriott TG, Kuchroo VK, Halliday WJ, Hellström I, Hellström KE: Monoclonal antiidiotypic antibodies related to a murine oncofetal bladder tumor antigen induce specific cell-mediated tumor immunity. *Proc Natl Acad Sci USA* 1985; 82:6286–6290.
18. Raychaudhuri S, Saeki Y, Chen JJ, Iribe H, Fuji H, Kohler H: Tumor-specific idiotype vaccines. II. Analysis of the tumor-related network response induced by the tumor and by internal image antigens (Ab2β). *J Immunol* 1987; 139:271–278.
19. Gaulton GN, Sharpe AH, Chang DW, Fields BN, Greene MI: Syngeneic monoclonal internal image anti-idiotopes as prophylactic vaccines. *J Immunol* 1986; 137:2930–2936.
20. Zhou EM, Chanh TC, Dreesman GR, Kanda P, Kennedy RC: Immune response to human immunodeficiency virus. *In vivo* administration of anti-idiotype induces an anti-gp160 response specific for a synthetic peptide. *J Immunol* 1987; 139:2950–2956.
21. Dalgleish AG, Thomson BJ, Chanh TC, Malkovsky M, Kennedy RC: Neutralisation of HIV isolates by anti-idiotypic antibodies which mimic the T4 (CD4) epitope: A potential AIDS vaccine. *Lancet* 1987; 2:1047–1050.
22. Kennedy RC, Eichberg JW, Lanford RE, Dreesman GR: Anti-idiotypic antibody vaccine for type B viral hepatitis in chimpanzees. *Science* 1986; 232:220–223.

23. Kennedy RC, Melnick JL, Dreesman GR: Antibody to hepatitis B virus induced by injecting antibodies to the idiotype. *Science* 1984; 223:930–931.
24. Schick MR, Dreesman GR, Kennedy RC: Induction of an anti-hepatitis B surface antigen response in mice by noninternal image (Ab2α) anti-idiotypic antibodies. *J Immunol* 1987; 138:3419–3425.
25. Thanavala YM, Brown SE, Howard CR, Roitt IM, Steward MW: A surrogate hepatitis B virus antigenic epitope represented by a synthetic peptide and an internal image anti-idiotypic antibody. *J Exp Med* 1986; 164:227–236.
26. Colucci G, Beazer Y, Waksal S: Interactions between HBV and polymeric human serum albumin. II. Development of syngeneic monoclonal anti-anti-idiotypes which mimic hepatitis B surface antigen in the induction of immune responsiveness. *Eur J Immunol* 1987; 17:371–374.
27. Keay S, Rasmussen L, Merigan TC: Syngeneic monoclonal anti-idiotype antibodies that bear the internal image of a human cytomegalovirus neutralization epitope. *J Immunol* 1988; 140:944–948.
28. Ertl HCJ, Finberg RW: Sendai virus-specific T-cell clones: Induction of cytolytic T-cells by an anti-idiotypic antibody directed against a helper T-cell clone. *Proc Natl Acad Sci USA* 1984; 81:2850–2854.
29. Ertl HCJ, Homans E, Tournas S, Finberg RW: Sendai virus-specific T cell clones. V. Induction of a virus-specific response by antiidiotypic antibodies directed against a T helper cell clone. *J Exp Med* 1984; 159:1778–1783.
30. Reagan KJ, Wunner WH, Wiktor TJ, Koprowski H: Anti-idiotypic antibodies induce neutralizing antibodies to rabies virus glycoprotein. *J Virol* 1983; 48:660–666.
31. Uytdehaag FGCM, Osterhaus ADME: Induction of neutralizing antibody in mice against poliovirus type II with monoclonal anti-idiotypic antibody. *J Immunol* 1985; 134:1225–1229.
32. Gell PGH, Moss PAH: Production of cell-mediated immune response to herpes simplex virus by immunization with anti-idiotypic heteroantisera. *J Gen Virol* 1985; 66:1801–1804.
33. Lathey JL, Courtney RJ, Rouse BT: Production, binding characteristics, and immunogenicity of heterologous anti-idiotypic antibody to herpes simplex virus glycoprotein C. *Viral Immunol* 1987; 1:13–24.
34. Mayer R, Ioannides C, Moran T, Johansson B, Bona C: Effect of syngeneic anti-idiotypic antibody on influenza virus neuraminidase antibody response. *Viral Immunol* 1987; 1:121–134.
35. Tanaka M, Sasaki N, Seto A: Induction of antibodies against Newcastle disease virus with syngeneic anti-idiotype antibodies in mice. *Microbiol Immunol* 1986; 30:323–331.
36. Paque RE, Miller R: Modulation of murine Coxsackievirus-induced myocarditis utilizing anti-idiotypes. *Viral Immunol* 1987; 1:207–224.
37. Ludwig DS, Finkelstein RA, Karu AE, Dallas WS, Ashby ER, Schoolnik GK: Anti-idiotypic antibodies as probes of protein active sites: Application to cholera toxin subunit B. *Proc Natl Acad Sci USA* 1987; 84:3673–3677.
38. Chanh TC, Huot RI, Schick MR, Hewetson JF: Anti-idiotypic antibodies against a monoclonal antibody specific for the trichothecene mycotoxin T-2. *Toxicol Appl Pharmacol* 1989; 100:201–207.
39. Hiernaux J, Bona C, Baker PJ: Neonatal treatment with low doses of anti-idiotypic leads to the expression of a silent clone. *J Exp Med* 1981; 153:1004–1008.

40. Rubinstein LJ, Yeh M, Bona CA: Idiotype-anti-idiotype network. II. Activation of silent clones by treatment at birth with idiotypes is associated with the expansion of idiotype-specific helper T cells. *J Exp Med* 1982; 156:506–521.
41. Marvel J, Tassignon J, Brait M, et al: The influence of V_k gene polymorphism on the induction of silent idiotypes in the arsonate system. *Mol Immunol* 1987; 24:463–469.
42. Francotte M, Urbain J: Induction of anti-tobacco mosaic virus antibodies in mice by rabbit antiidiotypic antibodies. *J Exp Med* 1984; 160:1485–1494.
43. Bona CA, Heber-Katz E, Paul WE: Idiotype-anti-idiotype regulation. Immunization with a levan-binding myeloma protein leads to the appearance of auto-anti-(anti-idiotype) antibodies and to the activation of silent clones. *J Exp Med* 1981; 153:951–967.
44. Gottlieb MS, Schroff R, Schanker HM, et al: *Pneumocystis carinii* pneumonia and mucosal candidiasis in previously healthy homosexual men. Evidence of a new acquired cellular immunodeficiency. *N Engl J Med* 1981; 305:1425–1431.
45. Masur H, Michelis MA, Greene JB, et al: An outbreak of community-acquired *pneumocystis carinii* penumonia. Initial manifestation of cellular immune dysfunction. *N Engl J Med* 1981; 305:1431–1438.
46. Siegal FP, Lopez C, Hammer GS, et al: Severe acquired immunodeficiency in male homosexuals, manifested by chronic perianal ulcerative herpes simplex lesions. *N Engl J Med* 1981; 305:1439–1444.
47. Wain-Hobson S, Sonigo P, Danos O, Cole S, Alizon M: Nucleotide sequence of the AIDS virus, LAV. *Cell* 1985; 40:9–17.
48. Ratner L, Haseltine W, Patarca R, et al: Complete nucleotide sequence of the AIDS virus, HTLV-III. *Nature* 1985; 313:277–284.
49. Muesing MA, Smith DH, Cabradilla CD, Benton CV, Laskey LA, Capon DJ: Nucleic acid structure and expression of the human AIDS/lymphadenopathy retrovirus. *Nature* 1985; 313:450–458.
50. Sanchez-Pescador R, Power MD, Barr PJ, et al: Nucleotide sequence and expression of an AIDS-associated retrovirus (ARV-2). *Science* 1985; 227:484–492.
51. Swain SL: T cell subsets and the recognition of MHC class. *Immunol Rev* 1983; 74:129–142.
52. Krensky AM, Reiss CS, Mier JW, Strominger JL, Burakoff SJ: Long-term human cytolytic T-cell lines allospecific for HLA-DR6 antigen are OKT4$^+$. *Proc Natl Acad Sci USA* 1982; 79:2365–2369.
53. Spits H, Borst J, Terhorst C, de Vries JE: The role of T cell differentiation markers in antigen-specific and lectin-dependent cellular cytotoxicity mediated by T8$^+$ and T4$^+$ human cytotoxic T cell clones directed at class I and class II MHC antigens. *J Immunol* 1982; 129:1563–1569.
54. Meuer SC, Schlossman SF, Reinherz EL: Clonal analysis of human cytotoxic T lymphocytes: T4$^+$ and T8$^+$ effector T cells recognize products of different major histocompatibility complex regions. *Proc Natl Acad Sci USA* 1982; 79:4395–4399.
55. Wilde DB, Marrack P, Kappler J, Dialynas DP, Fitch FW: Evidence implicating L3T4 in class II MHC antigen reactivity; monoclonal antibody GK1.5 (anti-L3T4) blocks class II MHC antigen-specific proliferation release of lymphokines, and binding by cloned murine helper T lymphocyte lines. *J Immunol* 1983; 131:2178–2183.

56. Swain SL, Dialynas DP, Fitch FW, English M: Monoclonal antibody to L3T4 blocks the function of T cells specific for class 2 major histocompatibility complex antigens. *J Immunol* 1984; 132:1118–1123.
57. Gay D, Maddon P, Sekaly R, et al: Functional interaction between human T-cell protein CD4 and the major histocompatibility complex HLA-DR antigen. *Nature* 1987; 328:626–629.
58. Doyle C, Strominger JL: Interaction between CD4 and class II MHC molecules mediates cell adhesion. *Nature* 1987; 330:256–259.
59. Reinherz EL, Schlossman SF: The differentiation and function of human T lymphocytes. *Cell* 1980; 19:821–827.
60. Dialynas DP, Quan ZS, Wall KA, et al: Characterization of the murine T cell surface molecule, designated L3T4, identified by monoclonal antibody GK1.5: Similarity of L3T4 to the human Leu-3/T4 molecule. *J Immunol* 1983; 131:2445–2451.
61. Marrack P, Endres R, Shimonkevitz R, et al: The major histocompatibility complex-restricted antigen receptor on T cells. II. Role of the L3T4 product. *J Exp Med* 1983; 158:1077–1091.
62. Maddon PJ, Littman DR, Godfrey M, Maddon DE, Chess L, Axel R: The isolation and nucleotide sequence of a cDNA encoding the T cell surface protein T4: A new member of the immunoglobulin gene family. *Cell* 1985; 42:93–104.
63. Maddon PJ, Molineaux SM, Maddon DE, et al: Structure and expression of the human and mouse T4 genes. *Proc Natl Acad Sci USA* 1987; 84:9155–9159.
64. Clayton LK, Hussey RE, Steinbrich R, Ramachandran H, Husain Y, Reinherz EL: Substitution of murine for human CD4 residues identifies amino acids critical for HIV-gp120 binding. *Nature* 1988; 335:363–366.
65. Classon BJ, Tsagaratos J, McKenzie IFC, Walker ID: Partial primary structure of the T4 antigens of mouse and sheep: Assignment of intrachain disulfide bonds. *Proc Natl Acad Sci USA* 1986; 83:4499–4503.
66. Clark SJ, Jefferies WA, Barclay AN, Gagnon J, Williams AF: Peptide and nucleotide sequences of rat CD4 (w3/25) antigen: Evidence for derivation from a structure with four immunoglobulin-related domains. *Proc Natl Acad Sci USA* 1987; 84:1649–1653.
67. Littman DR, Gettner SN: Unusual intron in the immunoglobulin domain of the newly isolated murine CD4 (L3T4) gene. *Nature* 1987; 325:453–455.
68. Benn S, Rutledge R, Folks T, et al: Genomic heterogeneity of AIDS retroviral isolates from North America and Zaire. *Science* 1985; 230:949–951.
69. Sonigo P, Alizon M, Staskus K, et al: Nucleotide sequence of the visna lentivirus. Relationship to the AIDS virus. *Cell* 1985; 42:369–382.
70. Myers G, Josephs SF, Berzofsky JA, Rabson AB, Smith TF, Wong-Staal F (eds): Human retroviruses and AIDS, 1989. A compilation and analysis of nucleic acid and amino acid sequences. Theoretical Biology and Biophysics, Los Alamos National Laboratory, Los Alamos, New Mexico, 1989.
71. Wong-Staal F, Shaw GM, Hahn BH, et al: Genomic diversity of human T-lymphotropic virus type III (HTLV-III). *Science* 1985; 229:759–762.
72. Alizon M, Wain-Hobson S, Montagnier L, Sonigo P: Genetic variability of the AIDS virus: Nucleotide sequence analysis of two isolates from African patients. *Cell* 1986; 46:63–74.

73. Weiss RA, Clapham PR, Weber JN, Dalgleish AG, Laskey LA, Berman PW: Variable and conserved neutralization antigens of human immunodeficiency virus. *Nature* 1986; 324:572–575.
74. Saag MS, Hahn BH, Gibbons J, et al: Extensive variation of human immunodeficiency virus type-1 *in vivo*. *Nature* 1988; 334:440–444.
75. Fisher AG, Ensoli B, Looney D, et al: Biologically divergent molecular variants within a single HIV-1 isolate. *Nature* 1988; 334:444–447.
76. Fauci AS: The human immunodeficiency virus: Infectivity and mechanisms of pathogenesis. *Science* 1988; 239:617–622.
77. Halstead SB, O'Rourke EJ: Dengue viruses and mononuclear phagocytes. I. Infection enhancement by non-neutralizing antibody. *J Exp Med* 1977; 146:201–217.
78. Robinson WE Jr, Montefiori DC, Mitchell WM: Antibody dependent enhancement of human immunodeficiency virus type 1 infection. *Lancet* 1988; 1:790–794.
79. Takeda A, Tuazon CU, Ennis FA: Antibody-enhanced infection by HIV-1 via Fc receptor-mediated entry. *Science* 1988; 242:580–583.
80. Dalgleish AG: Human trials of AIDS vaccines: Novel means of passive and active immunotherapy. *AIDS 2* 1988; (suppl 1):S129–S131.
81. Snart RS: Human trials of experimental AIDS vaccines: Recombinant envelope proteins. *AIDS 2* 1988; (suppl 1):S107–111.
82. Kennedy RC, Chanh TC: Perspectives on developing anti-idiotype based vaccines for controlling HIV infection. *AIDS 2* 1988; (suppl 1):S119–S127.
83. Smith DH, Byrn RA, Marsters SA, Gregory T, Groopman JE, Capon DJ: Blocking of HIV-1 infectivity by a soluble, secreted form of the CD4 antigen. *Science* 1987; 238:1704–1707.
84. Fisher RA, Bertonis JM, Meier W, et al: HIV infection is blocked *in vitro* by recombinant soluble CD4. *Nature* 1988; 331:76–78.
85. Hussey RE, Richardson NE, Kowalski M, et al: A soluble CD4 protein selectively inhibits HIV replication and syncytium formation. *Nature* 1988; 331:78–81.
86. Deen KC, McDougal JS, Inacker R, et al: A soluble form of CD4 (T4) protein inhibits AIDS virus infection. *Nature* 1988; 331:82–84.
87. Traunecker A, Luke W, Karjalainen K: Soluble CD4 molecules neutralize human immunodeficiency virus type 1. *Nature* 1988; 331:84–86.
88. Sattentau QJ, Dalgleish AG, Weiss RA, Beverley PCL: Epitopes of the CD4 antigen and HIV infection. *Science* 1986; 234:1120–1123.
89. Chanh TC, Dreesman GR, Kennedy RC: Monoclonal anti-idiotypic antibody mimics the CD4 receptor and binds human immunodeficiency virus. *Proc Natl Acad Sci USA* 1987; 84:3891–3895.
90. Beasley RP: Hepatitis B virus as the etiologic agent in hepato-cellular carcinoma: Epidemiologic considerations. *Hepatology* 1982; 2:21–26.
91. McAuliffe VJ, Purcell RH, Gerin JL: Type B hepatitis: A review of current prospects for a safe and effective vaccine. *Rev Infect Dis* 1980; 2:470–492.
92. Levene C, Blumberg BS: Additional specificities of Australia antigen and the possible identification of hepatitis carriers. *Nature* 1969; 221:195–196.
93. Le Bouvier GL: The heterogeneity of Australia antigen. *J Infect Dis* 1971; 123:671–675.
94. Bancroft WH, Mundon FK, Russell PK: Detection of additional antigenic determinants of hepatitis B antigen. *J Immunol* 1972; 109:842–848.

95. Szmuness W, Stevens CE, Harley EJ, et al: Hepatitis B vaccine. Demonstration of efficacy in a controlled clinical trial in a high-risk population in the United States. *N Engl J Med* 1980; 303:833–841.
96. Kennedy RC, Dreesman GR: Common idiotypic determinant associated with human antibodies to hepatitis B surface antigen. *J Immunol* 1983; 130:385–389.
97. Kennedy RC, Sanchez Y, Ionescu-Matiu I, Melnick JL, Dreesman GR: A common human anti-hepatitis B surface antigen idiotype is associated with the group a conformation-dependent antigenic determinant. *Virology* 1982; 122:219.
98. Kennedy RC, Dreesman GR, Sparrow JT, et al: Inhibition of a common human anti-hepatitis B surface antigen idiotype by a cyclic synthetic peptide. *J Virol* 1983; 46:653–655.
99. Kennedy RC, Ionescu-Matiu I, Sanchez Y, Dreesman GR: Detection of interspecies indiotypic cross-reactions associated with antibodies to hepatitis B surface antigen. *Eur J Immuno* 1983; 13:232–235.
100. Kennedy RC, Adler-Storthz K, Henkel RD, Sanchez Y, Melnick JL, Dreesman GR: Immune response to hepatitis B surface antigen: Enhancement by prior injection of antibodies to the idiotype. *Science* 1983; 221:853–855.
101. Kennedy RC, Sparrow JT, Sanchez Y, Melnick JL, Dreesman GR: Enhancement of viral hepatitis B antibody (anti-HBS) response to a synthetic cyclic peptide by priming with anti-idiotype antibodies. *Virology* 1984; 136:247–252.
102. Kennedy RC, Dreesman GR: Enhancement of the immune response to hepatitis B surface antigen. *In vivo* administration to antiidiotype induces anti-HB that expresses a similar idiotype. *J Exp Med* 1984; 159:655–665.
103. Kennedy RC, Eichberg JW, Dreesman GR: Lack of genetic restriction by i potential anti-idiotype vaccine type B viral hepatitis. *Virology* 1986; 148:369–374.
104. Rajewsky K, Takemori T: Genetics, expression, and function of idiotypes. *Ann Rev Immunol* 1983; 1:569–607.
105. Peltola H, Kayhty H, Sivonen A, Makela PH: *Haemophilus influenzae* type B capsular polysaccharide vaccine in children: A double-blind field study of 100,000 vaccines 3 months to 5 years of age in Finland. *Pediatrics* 1977; 60:730–737.
106. Peltola H, Kayhty H, Virtanen M, Makela PH: Prevention of *Haemophilus influenzae* type B bacteremic infections with the capsular polysaccharide vaccine. *N Engl J Med* 1984; 310:1561–1566.

CHAPTER 8

Utilization of Anti-Idiotypic Antibodies as Molecular Probes of Virus–Receptor Interaction

David B. Weiner, Daniel E. McCallus, William V. Williams, and Mark I. Greene

Knowledge of the molecular basis of ligand receptor interactions may lead to strategies for the design of substances that can prevent viral-mediated pathology. We will describe our work in two different viral systems. First we review our studies of the reovirus type 3 system. We have utilized the reovirus type 3–cellular receptor interaction as a model system to develop such a strategy. We demonstrate the utility of monoclonal anti-idiotypic (anti-Id) antibodies as starting points for the rational design of peptides that mimic the specificities of the receptor-binding structures of viruses. In this model system we were able to specifically determine the receptor interaction sites by sequence homology with the viral hemagglutinin. Based on these sequences we produced functionally active mimics that interfere with virus binding and maintain, when suitably constructed, many properties of the hemagglutinin.

Second we studied certain aspects of the biology of a human retrovirus. Here we applied anti-idiotype technology in a novel way to elucidate integral parts of the pathway of AIDS virus entry. We present data defining other molecules in addition to the CD4 receptor protein that are involved in post-binding entry events mediated by the retroviral envelope glycoproteins.

The Reovirus Type 3 System

Reovirus type 3 binds to specific cellular receptors by the hemagglutinin molecule (reovirus type 3 hemagglutinin or HA3) (1, 2). This interaction determines both the tissue tropism of viral infection and also influences cellular metabolism (1–13). As a specific probe for the reovirus type 3 receptor (Reo3R), Reo3R-binding antibodies were developed utilizing an anti-Id approach (10, 14, 15). Initially, xenogeneic, polyclonal anti-Reo3R, anti-Id antibodies were produced (10, 15). Antibodies (anti-idiotypes) were screened by radioimmunoassay (RIA) for their ability to inhibit the binding of labeled anti-HA3–specific antibodies to the viral HA3 (10, 15). Subsequently, monoclonal antibodies (MAb) directed against HA3 were developed (16, 17) and screened for the presence of internal image idiotype by their ability to bind xenogeneic anti-idiotype (15). One MAb, termed 9BG5, a type 3-specific, HA3-binding, neutralizing antibody, strongly bound rabbit anti-idiotype (10, 15). The idiotype displayed by 9BG5 and recognized by type 3-specific, rabbit anti-idiotype is termed ID3, and the corresponding anti-idiotype is termed anti-Id3.

More recently monoclonal anti-Id3s have been constructed following the immunization of syngeneic mice with 9BG5 hybridoma cells (10, 18, 19). The MAb anti-Id3s were screened for their ability to inhibit the binding of 9BG5 to purified HA3 protein as described above. One MAb that blocked binding was designated 87.92.6 Binding of 87.92.6 to a panel of cells mirrored that of reovirus type 3 (18). Direct evidence that MAb anti-Id3 recognizes reovirus type 3 receptor was achieved by the demonstration that prior incubation with anti-Id3 specifically inhibited the binding of type 3 virus to R1.1 lymphoma cells by 90%. The 87.92.6 also mimicked the HA3 functionally by inhibiting DNA synthesis of cells upon Reo3R cross-linking (13) and by being able to induce both cellular and humoral immunity to reovirus type 3 (20–22). Due to these properties, and an ability to bind cellular receptors (Reo3R), 87.92.6 was considered an internal image, anti-Id antibody.

We have analyzed the specific interactions of both reovirus type 3 and 87.92.6 with the Reo3R. The heavy and light chain variable region genes (V_H and V_L, respectively) of 87.92.6 have been sequenced, and compared to the deduced amino acid sequences to the reovirus type 3 hemagglutinin (HA3) sequence (23). These studies revealed sequence similarities between amino acids 317-332 of HA3 and a combined determinant comprised of the V_H CDR II and the V_L CDR II of 87.92.6. Peptides corresponding to the amino acid sequences displaying such homologies were synthesized. They are specifically denoted VH peptide, V_L peptide, and reo peptide. In a series of studies (24–28), it was demonstrated that V_L peptide is bound by the neutralizing anti-reovirus type 3 MAb (9B.G5), and that V_L peptide inhibits the interaction of 9B.G5 with both reovirus type 3 and 87.92.6. The V_L peptide also inhibits the interaction of reovirus type 3 and 87.92.6 with the reovirus type 3 receptor. This indicates that the amino acid sequence shared by V_L peptide and the HA3 (amino acids 323 to 332 of the HA3) define the cell-attachment site/neutralizing-antibody site of reovirus type 3.

Perturbation of the reovirus type 3 receptor of a variety of cell lines disrupts the cell cycle as demonstrated by the inhibition of cellular DNA synthesis (11, 12). The peptides described above have been utilized to investigate these effects further. We determined that V_L peptide (with an added amino terminal cysteine residue to promote ligand dimerization) was able to inhibit proliferation of a variety of cell lines. Furthermore, dimerized peptide is able to inhibit concanavalin A-dependent lymphocyte proliferation (28–31). This analog of the V_L peptide also down-modulates the Reo3R on murine thymoma (R1.1) cells. Monomeric V_L peptide, without the added cysteine residue, is unable to induce this effect.

This study demonstrates an important feature of the reovirus type 3 receptor function: cross-linking of the receptor is required to manifest its cellular effects. These studies indicate that receptor binding and aggregation alone are able to initiate events culminating in inhibition of DNA synthesis.

Development of Molecular Models of the Viral Cell-Attachment Site

We have modeled the three-dimensional structure of the receptor interaction sites of reovirus type 3 and its internal image (24–28). Our studies utilized a comparative modeling approach based on the known structures of other immunoglobulin CDR IIs with similar sequences. The CDR II with a defined structure provided starting geometries upon which model structures for the HA3 and V_L peptide sites could be developed. The preliminary structures were then subjected to energy minimization calculations to produce the final working models (24–28). Schematic diagrams of these model receptor interaction sites are shown in Figure 8.1.

Several structures have been proposed to serve as reovirus receptors on cells (32–35). Among these, sialic acid has been implicated as a component of the reovirus receptor on some cells (34, 36). Since V_L peptide inhibits binding of the HA3 to these cells, it is likely that there is a direct interaction between V_L peptide and sialic acid on these cells. This interaction inhibits the subsequent attachment of the HA3 to sialic acid.

We have observed a critical interaction between the hydroxyl groups in V_L peptide and the Reo3R on murine L cells in determining binding (29, 37). This has led to the development of molecular models of potential interactions

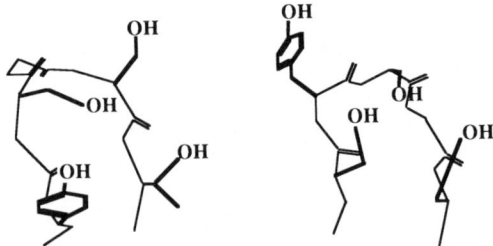

Fig. 8.1. Schematic diagram of the 87.92.6 VL CDR II (left) and the reovirus type 3 receptor-interaction epitope (right). The model structures were developed by a comparative approach as described (25–29). Schematized here are the reverse-turn regions of the putative receptor interaction sites. The amino acids in the turn regions depicted are:

87.92.6 VL CDR II	Tyr–Ser–Gly–Ser–Thr
REOVIRUS HA3 EPITOPE	Ser–Tyr–Ser–Gly–Ser

Prior modeling studies (26) indicate that the side chains of the amino acids in the reverse turns can be superimposed when oriented at a 90° angle to each other, which indicates similar potential binding orientations of the side chains.

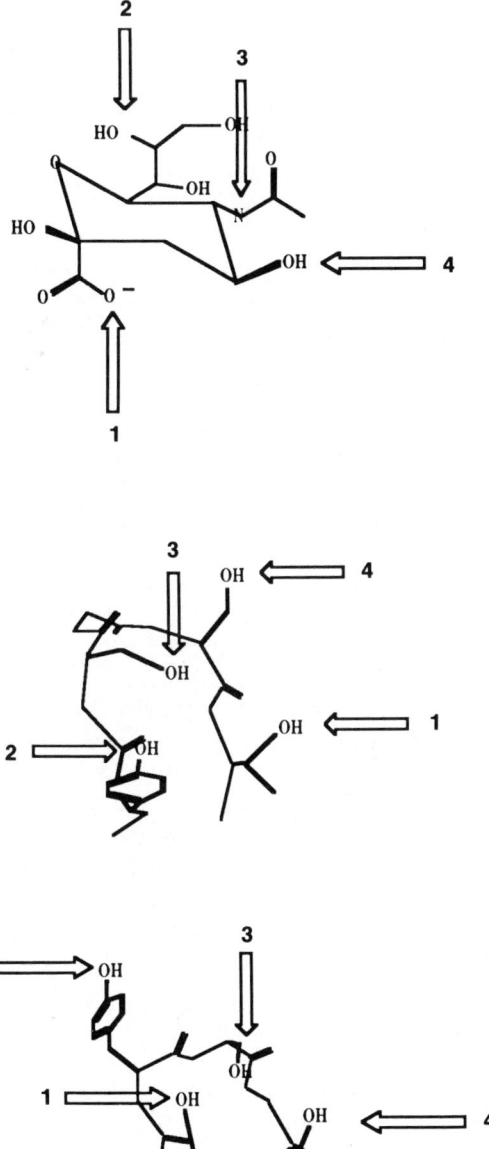

Fig. 8.2. Schematic diagram of predicted interaction sites between sialic acid/Reo3R (top) and 87.92.6 VL CDR II (middle)/HA3 (bottom). Molecular modeling studies (30, 38) indicate hydrogen bond formation between the indicated side chains of sialic acid with the corresponding numbered side chains of 87.92.6 VL CDR II and the reovirus type 3 receptor-interaction site.

involving V_L peptide and sialic acid, and the HA3 determinant and sialic acid. To develop these models, starting geometries for the V_L peptide and HA3 epitope were utilized as previously described (25). The conformation of sialic acid atoms interacting with the influenza virus hemagglutinin (IHA) (36) was used for modeling. Hydrogen-bonding interactions identified in the IHA–sialic acid interaction (which involves residues distant in sequence) are maintained by the V_L peptide and HA3 epitopes. After appropriate starting geometries of the interaction models were obtained, energy minimization was carried out utilizing the program DISCOVER (BIOSYM Technologies, California). The resultant structures indicate that four hydrogen bond interactions are involved in the binding of the HA3 and V_L peptide epitopes to sialic acid (37). A schematic of these potential interactions is shown in Figure 8.2.

These studies indicate that anti-idiotypes can be utilized as molecular mimics to develop an understanding of the structural interactions important for receptor binding. Based on these detailed studies in the reovirus system, we have approached studies of human retrovirus binding and entry utilizing anti-Id antibodies.

The AIDS/HIV-1 Retrovirus

Human immunodeficency virus (HIV-1) is the etiological agent of the acquired immunodeficency syndrome (AIDS) and a spectrum of related disorders (38–40). The clinical manifestations of this disease include immune system failure and dementia. Immunological capacity is lost in a progressive and irreversible manner. Approximately 1 million antibody-positive, virally exposed individuals may eventually progress to the active disease state.

A single species of retrovirus is the causative agent in AIDS, HIV-1—originally named human T cell lymphotropic virus (HTLV-III), AIDS-associated retrovirus (ARV), or lymphadenopathy associated virus (LAV). The AIDS virus is most closely associated with the lentivirus family of retroviruses. These diploid, single-stranded RNA viruses induce progressive and fatal disease in their hosts.

HIV-1 is similar in its genomic and overall structural organization to other previously characterized retroviruses (38–40). The HIV virions are composed of dimers of single-stranded RNA that are organized around gene products of the *gag* region. These are assembled into an electron-dense cylindrical core. The core is surrounded by a lipid bilayer gleaned from the host cell as the virus buds from the infected cell surface. Two viral envelope glycoproteins, one entirely external (gp120) and one membrane spanning (gp41) are displayed on the outside of the virus lipid bilayer (41). These two viral proteins stimulate the host's immune response, which includes active cellular and humoral immunity (42). The gp120 envelope glycoprotein specifically interacts with the cellular attachment receptor for the virus.

Receptor Interactions

Disease manifestations are a function of the cellular tropism of the virus for helper T cells, macrophages and histocytic cell lineages. Viral selectivity is determined by the interaction of the envelope glycoproteins of HIV-1, with the CD4 antigen (43, 44). The human CD4 molecule is a 55-kDa glycoprotein involved in interaction with class II major histocompatibility structures. Tissue surveys have demonstrated that CD4 is expressed uniquely on T cells and antigen-presenting cells, as well as rare, B cell tumor lines (45). HIV-1 tropism results from interactions between the virus envelope gp120 and a high-affinity binding site on the CD4 glycoprotein that permit viral adsorption (46-49). Both the specific interaction sites on human CD4 and HIV-1 gp120 have been determined. Site-directed mutagenesis studies suggest that three separate regions of the virus envelope, spanning AA363, AA419, and AA473, fold to form the CD4-binding site (41). Peptide inhibition studies using HIV-1 envelope analogs suggest a binding site in the same general 3' region of the virus and also an area of interaction at aa 397-439 (50). It is likely that the viral epitope involved in CD4 interactions is nonlinear.

In contrast to the gp120 epitope for CD4, the CD4 attachment site is a linear epitope composed of amino acid residues within the first cysteine-bonded loop of the deduced sequence of CD4. Peptide mapping, MAb mapping, and site-directed mutagenesis, as well as molecular constructs revealed the minimal epitopes necessary for interactions with gp120 (46-49). The sites necessary for interaction span AA 30-60 with the amino acid residues between AA 45-55 representing the major region detected in most studies.

After the initial attachment of virus to the human cell surface CD4 molecule, other regions of the virus envelope are believed to be important in initiating fusion between the viral and cellular membranes (50-53). This secondary interaction precedes viral entry, uncoating, and replication. The postbinding requirements for virus entry are poorly understood.

Experimental Approaches to HIV Biology

Recent evidence suggests that HIV entry into cells is a complex process. For example, CD4 expression on murine cells does not allow infection of the murine cells, even though these CD4 molecules bind HIV-1 in a manner indistinguishable from CD4 on human cells (45, 52, 54). Recent studies demonstrate that CD4 expressed on the murine cellular background functions identically to CD4 molecules expressed on human cell lines. The block to infection does not appear to reside within the cellular machinery of the murine cell.

We and others have demonstrated that direct electroporation of HIV-1 proviral clones into murine cells leads to HIV-associated reverse transcriptase activity and production of infectious virus (52). In addition introduction of

HIV provirus into transgenic mice generates a disease pattern that resembles that observed in human infection.

It is possible that the block to infection of murine cells was due to suppression of HIV in the murine background. To address this issue, we analyzed HIV infection in constructed interspecific mouse × human hybridomas (52) (Fig. 8.3). A subset of interspecific hybridomas we transfected with plasmids containing human CD4 cDNA demonstrated productive infection and evidence of syncytia formation. Neither the murine parent cell lines nor the human parental cell lines in the absence of human CD4 expression supported HIV infection. These results demonstrate that infection in the murine background can occur if both CD4 and additional components are

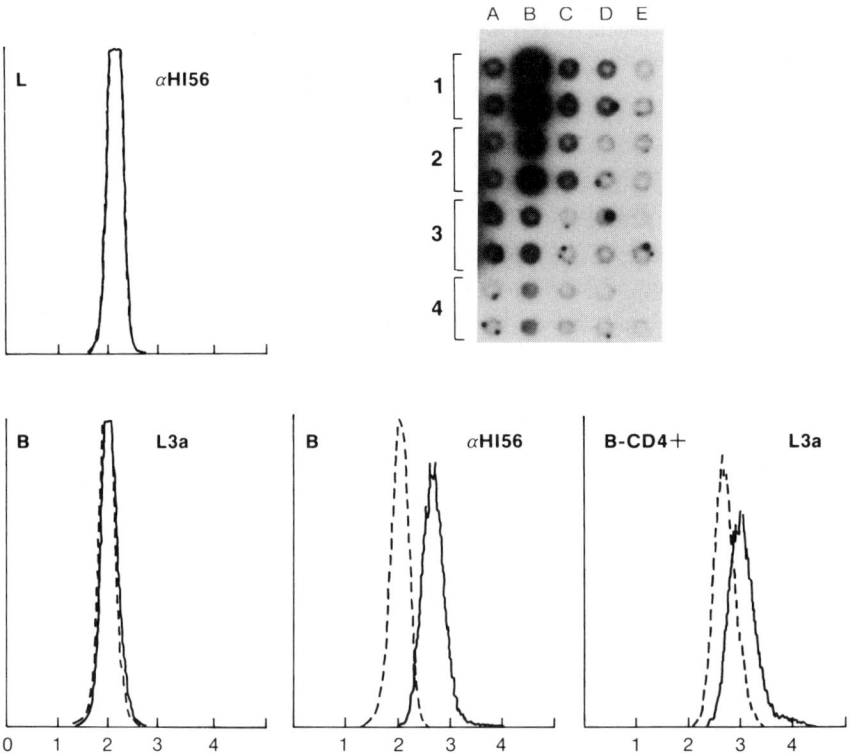

Fig. 8.3. Interspecific hybridoma studies. The flow microfluorimetry reactivity patterns for murine cells (L) or untransfected interspecific hybridomas cells (B) or interspecific hybridoma cells transfected with a CD4-expression vector (B–CD4+) are presented. The murine cells are not reactive with anti-H156 sera (see text); the interspecific hybridoma line displays reactivity with both anti-H156 and, after transfection with CD4 construct, displays this viral receptor. *Inset:* The infection profile of several hybridomas is presented as dilutions of reverse transcriptase activity from infected cultures. The hybridoma cell line B is clearly infectible.

Table 8.1. Syncytia-inhibiting activity.[a,b] Inhibition of syncytia formation by H156 sera and serum fractions. The IgM and IgG fractions were purified as previously described (Vaccines 89). Sup T1 cells were preincubated with or without the indicated sera or fractions for 30 minutes, at which time virus-producing cells (WMJ) were added; incubation proceeded for an additional 24 hours before the plates were scored for the degree of syncytia observed (10).

	1/8	1/16	1/32	1/64	1/128	1/256
Protein A purified	0	0	0	0	0	1S
IgM fraction	0	0	2M	1M	2M	2M
△H156 sera	0	0	0	0	1M	2M
Normal human sera	4L	4L	4L	4L	4L	4L

[a]Degree of syncytia formation: 4, full; 0, none.
[b]Size of syncytia formed: S, small; M, medium; L, large.

supplied at the same time (55). Similarily, CD4-independant infection of human cells has recently been reported (53). Together, the results imply that surface molecules in addition to CD4 are important in mediating HIV entry into target cell lines. We utilized anti-Id technology to define these accessory molecules.

One approach to define ancillary cellular components active in syncytia formation and viral internalization is to screen patient antisera for reactivities that will inhibit virus interactions at the cell surface. Some AIDS patient antisera occasionally contain low levels of syncytia-inhibiting antibodies (51, 54). We screened over 300 HIV+ patients' sera to study this subset of patients. These antisera possessed reactivities with viral envelope determinants other than those important for CD4-gp120 interactions. By using syncytia assays, an asymptomatic HIV-1-infected individual was identified whose sera (sera H156) was highly reactive to HIV-1 proteins; the sera also possessed potent anti-syncytia activity (51) (Table 8.1). Fractionation of the sera demonstrated that the highest concentration of antisyncytia activity was contained within the IgG fraction. The antisyncytial antibodies inhibited several isolates including the WMJ and IIIb isolates of HIV-1.

We assayed the ability of patient sera to inhibit CD4/gp120 glycoprotein interactions directly in immunoprecipitation studies. Antibodies directed at the CD4 molecule that inhibited syncytia formation also inhibited gp120-mediated coimmunoprecipitation of CD4 (Fig. 8.4). However, the H156 antisera inhibited syncytia formation without inhibiting the coimmunoprecipitation of CD4. This indicated that the H156 antibodies were inhibiting HIV-1-mediated syncytia formation by some other mechanism and not by inhibition of the CD4-gp120 interaction. This interaction was mediated

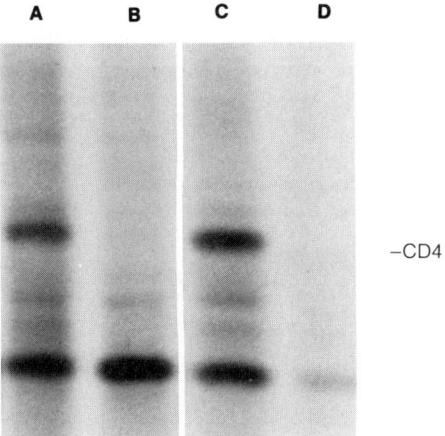

Fig. 8.4. H156 interaction with gp120. The ability of patient sera with antisyncytia activity to block a gp120–CD4 interaction was assayed by immunoprecipitation analysis. Radiolabeled CD4 was added to radiolabeled, virally infected cell lysates and immunoprecipitated by H156 sera (A), H156 sera and the leu 3a monoclonal antibody that blocks CD4–gp120 interactions (B), H156 sera and an irrelevant antibody that does not block CD4–gp120 interactions (C), and control sera (D). H156 does not appear to inhibit virus–cellular receptor interactions.

through binding of regions of the envelope of HIV at a region distinct from the CD4 interaction site. This site may be important in a postbinding entry-related interaction.

To further characterize the H156 sera reactivity, as well as to identify possible host cell molecules interacting with virus envelope glycoproteins, anti-Id antibodies were constructed (51) to the protein A purified H156 immunoglobulins. Mice were immunized subcutaneously with purified immunoglobulin at twice weekly intervals. Their immune responses were monitored after six immunizations. Radioimmunoassay (RIA) analysis of the anti-Id antibody absorbed with normal human immunoglobulins demonstrated specific responses against the immunizing immunoglobulins with minimal binding to normal human immunoglobulin (51). The anti-Id antisera (anti-H156) demonstrated much greater idiotype-specific binding activity to H156 then to pooled AIDS patients' immunoglobulin.

The anti-H156 antibodies were used to identify determinants important in postbinding HIV entry-mediated events. Flow microfluorimetry analysis demonstrated anti-H156 reactivity with human cells that lacked CD4 expression. We observed comparable anti-H156 binding with both CD4+ and CD4− human cell lines, which demonstrated that the major determinant(s) recognized by the anti-Id antibodies could not be the CD4 glycoprotein. This is illustrated in Figure 8.5, which shows that the anti-H156 sera demonstrated minimal reactivity with murine cells, even if those cells were

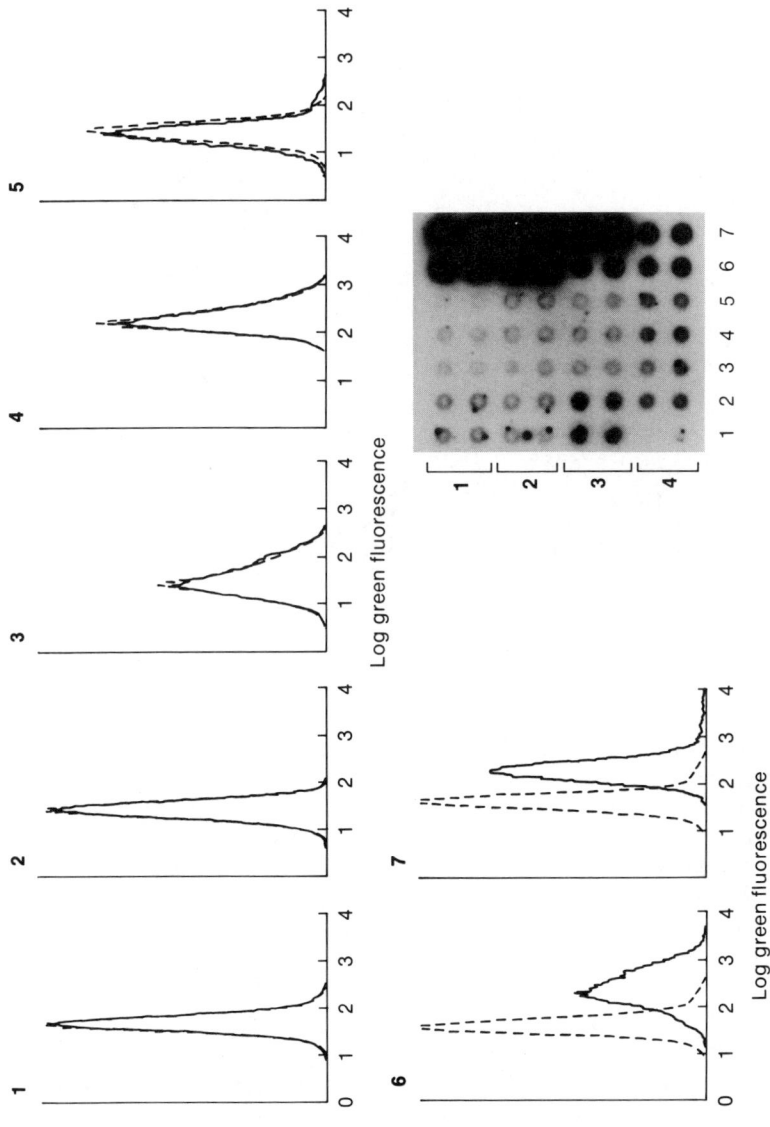

Fig. 8.5. Reactivity of anti-H156 with cell lines. Murine cell line (1–5) or two human cell lines (6–7) were stained with anti-H156 sera and analyzed by flow microfluorimetry for their reactivity patterns. The human cell lines are clearly reactive in this assay. *Inset*: Infectivity profiles of CD4+ constructs of the murine (1–5) and human (6–7) cell lines as analyzed by dilutions of reverse transcriptase activity are presented. Cell lines 6 and 7 display significant levels of reverse transcriptase activity upon HIV infection. The murine cells remain refractory to HIV infection.

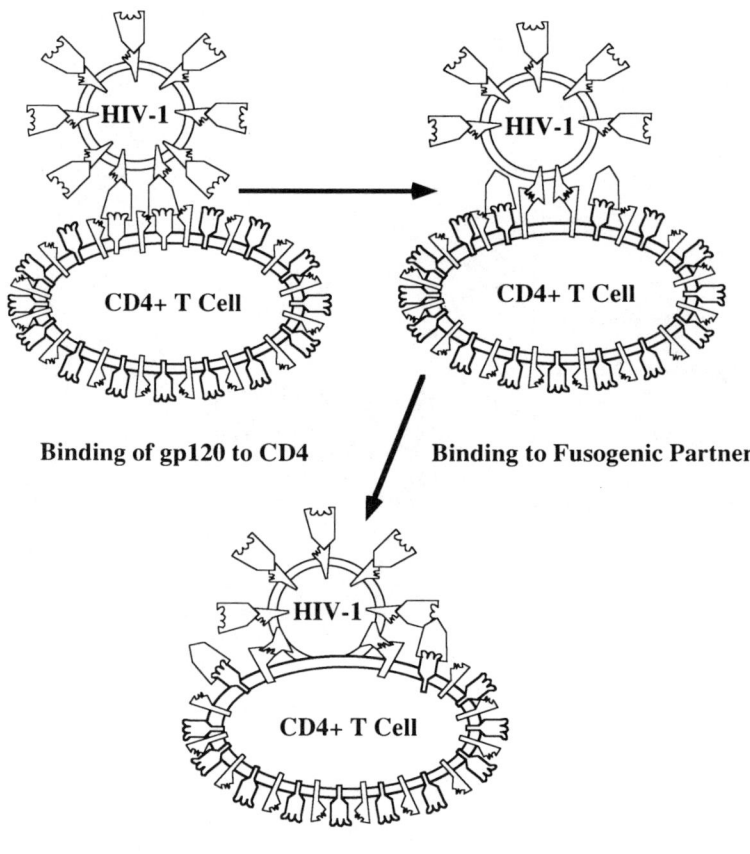

Fig. 8.6. Steps in cell entry of HIV-1. The HIV-1 virus interacts with its high-affinity cellular receptor, CD4, through the gp120 external envelope glycoprotein. After binding, a presumed conformational change in the structure of the gp120–gp41 envelope complex facilitates subsequent interaction with a non-CD4, ancillary, host cell surface molecule(s). This molecule is present on a variety of CD4− human lines. This specific interaction facilitates viral entry into target cells.

transfected with CD4 expression vectors. Human tumor cell lines, however, reacted significantly with anti-H156. These cell lines, when transfected with CD4 expression vectors, demonstrated an infectivity pattern consistent with susceptibility to HIV infection. Murine cells did not react with anti-H156 and were not susceptible to HIV infection. Furthermore, analysis of the interspecific hybridoma cell lines, which specifically supported HIV infection, demonstrated the presence of both anti-CD4 and anti-H156 reactivities (Fig. 8.3). These data suggest that we have identified a surface structure ancillary to the

CD4 receptor that may be important in HIV-1 pathogenesis. Based on these studies, a model for HIV-entry events is presented (Fig. 8.6).

In recent studies, anti-H156 was found to immunoprecipitate several polypeptides including a band at 25 to 35 kD, as well as bands at 45 to 60, 80 to 100, and 180 to 220 kD. Although further biochemical characterization of the relevant fusion moieties is necessary, it is likely that this anti-Id antibody may be useful in the further characterization of the pathogenesis of HIV-1, as well as in the development of novel therapeutics aimed at controlling viral infections that require membrane fusion.

Strategies involving anti-Id, antireceptor antibodies that are aimed at defining features of and controlling HIV infection are novel. Significant progress has been made in eliciting immune responses to the virus and in understanding HIV biology. The construction of new, high-titered reagents will have a significant impact on potential anti-HIV strategies exploiting the idiotypic network.

References

1. Epstein RL, Powers ML, Rogart RB, Weiner HL: Binding of ^{125}I-labeled reovirus to cell surface receptors. *Virology* 1984; 133:46.
2. Lee PWK, Hayes EC, Joklik WK: Protein σ1 is the reovirus cell attachment protein. *Virology* 1981; 108:156.
3. Joklik WK: *The Reoviridae*. New York: Plenum Press; 1983: pp1–78.
4. Weiner HL, Powers ML, Fields BN: Absolute linkage of virulence and central nervous system cell tropism of reoviruses to viral hemagglutinin. *J Infect-Dis* 1980; 141:609.
5. Weiner HL, Drayna D, Averill DR Jr, Fields BN: Molecular basis of reovirus virulence: Role of the S1 gene. *Proc Natl Acad Sci USA* 1977; 74:5744.
6. Kaye KM, Spriggs DR, Bassel-Duby R, Fields BN, Tyler KL: Genetic basis for altered pathogenesis of an immune-selected antigenic variant of reovirus type 3 (Dearing). *J Virol* 1986; 59:90.
7. Kilham L, Margolis G: Hydrocephalus in hamsters, ferrets, rats and mice following inoculation with reovirus type I. I. Virologic studies. *Lab Invest* 1969; 21:183.
8. Raine CS, Fields BN: Reovirus type encephalitis—a virologic and ultrastructural study. *J Neuropathol Exp Neurol* 1973; 32:19.
9. Maratos-Flier E, Kahn CR, Spriggs DR, Fields BN: Ammonium inhibits cytotoxicity of reovirus, a nonenveloped virus. *J Clin Invest* 1982; 72:617.
10. Nepom JT, Tarideau M, Epstein RL, et al: Virus binding receptors. Similarities to immune receptors as determined by anti-idiotypic antibodies. *Surv Immunol Res* 1982; 1:255.
11. Gomatos PJ, Tamm I: Macromolecular synthesis in reovirus-infected cells. *Biochem Immunol Biophys Acta* 1963; 72:651.
12. Sharpe AH, Fields BN: Reovirus inhibition of cellular DNA synthesis: Role of the S1 gene. *J Virol* 1981; 38:389.
13. Gaulton GN, Greene MI: Inhibition of cellular DNA synthesis by reovirus occurs through a receptor-linked signaling pathway that is mimicked by antireceptor antibody. *J Exp Med* 1989; 169:197.

14. Jerne NK: Towards a network theory of the immune system. *Ann Immunol (Paris)* 1974; 125:373.
15. Nepom JT, Weiner HL, Dichter MA, et al: Identification of a hemagglutinin-specific idiotype associated with reovirus recognition shared by lymphoid and neural cells. *J Exp Med* 1982; 155:155.
16. Finberg R, Spriggs DR, Fields BN: Host immune response to reovirus: CTL recognize the major neutralization domain of the viral hemagglutinin. *J Immunol* 1982; 129:2235.
17. Burstin SJ, Spriggs DR, Fields BN: Evidence for functional domains on the reovirus type 3 hemagglutinin. *Virology* 1982; 117:146.
18. Noseworthy JH, Fields BN, Dichter MA, et al: Cell receptors for the mammalian reovirus. I. Syngeneic monoclonal anti-idiotypic antibody identifies a cell surface receptor for reovirus. *J Immunol* 1983; 131:2533.
19. Kauffman RS, Noseworthy JH, Nepom JT, Finberg R, Fields BN, Greene MI: Cell receptors for mammalian reovirus. II. Monoclonal anti-idiotypic antibody blocks viral binding to cells. *J Immunol* 1983; 131:2539.
20. Sharpe AH, Gaulton GN, McDade KK, Fields BN, Greene MI: Syngeneic monoclonal antiidiotype can induce cellular immunity to reovirus. *J Exp Med* 1984; 160:1195.
21. Sharpe AH, Gaulton GN, Ertl HCJ, et al: Cell receptors for the mammalian reovirus. IV. Reovirus-specific cytolytic T cell lines that have idiotypic receptors recognize anti-idiotypic B cell hybridomas. *J Immunol* 1985; 134:2702.
22. Gaulton GN, Sharpe AH, Chang DW, Fields BN, Greene MI: Syngeneic monoclonal internal image anti-idiotopes as prophylactic vaccines. *J Immunol* 1986; 137:2930.
23. Bruck C, Co MS, Slaoui M, et al: Nucleic acid sequence of an internal image-bearing monoclonal anti-idiotype and its comparison to the sequence of the external antigen. *Proc Natl Acad Sci USA* 1986; 83:6578.
24. Williams WV, Guy HR, Weiner DB, Rubin D, Greene MI: Structure of the neutralizing epitope of the reovirus type 3 hemagglutinin. *Vaccines 88*. New York, Cold Spring Harbor Press, 1988, pp 25–28.
25. Williams WV, Guy HR, Rubin DH, et al: Sequence of the cell-attachment sites of reovirus type 3 and its anti-idiotypic/antireceptor antibody: Modeling of their three-dimensional structures. *Proc Natl Acad Sci USA* 1988; 85:6488.
26. Cohen JA, Williams WV, Greene MI: Molecular aspects of reovirus-host cell interactions. *Microbiol Sci* 1988; 5:265–270.
27. Williams WV, Weiner DB, Guy HR, Greene MI: Molecular basis for internal image mimicry of the reovirus type 3 cell attachment site. *Ann d'Immuno (Institut Pasteur)* 1988; 139:659–675.
28. Williams WV, Guy HR, Greene MI: Three-dimensional structure of a functional internal image. *Viral Immunol.*
29. Williams WV, Kieber-Emmons T, Weiner DB, Greene MI: Molecular analysis of a ligand–receptor interaction utilizing antibody structure. Proceedings, International Conference on Idiotypes.
30. Williams WV, Moss DA, Weiner DB, Cohen JA, Guy HR, Greene MI: Anti-idiotype modeled peptides with biologic activity. In: Haddon JW, Spreafico F, Yamamura Y, Austen KF, Dukor P, Masek K, eds: *Advances in Immunopharmacology, 4.* New York: Pergamon Press; 1988: pp 119–126.

31. Williams WV, Moss DA, Kieber-Emmons T, et al: Development of biologically active peptides based on antibody structure. *Proc Natl Acad Sci USA* 1989; 86:5537.
32. Co M-S, Gaulton GN, Fields BN, Greene MI: Isolation and biochemical characterization of the mammalian reovirus- type 3 cell-surface receptor. *Proc Natl Acad Sci USA* 1985; 82:1494.
33. Co M-S, Gaulton GN, Tominaga A, Homcy CH, Fields BN, Greene MI: Structural similarities between the mammalian β-adrenergic receptor and reovirus type 3 receptors. *Proc Natl Acad Sci USA* 1985; 82:5315.
34. Gentsch RJ, Pacitti AF: Effect of neuraminidase treatment of cells and effect of soluble glycoproteins on type 3 reovirus attachment to murine L cells. *J Virol* 1985; 56:356.
35. Pacitti AF, Gentsch JR: Inhibition of reovirus type 3 binding to host cells by sialylated glycoproteins is mediated through the viral cell attachment protein. *J Virol* 1987; 61:1407.
36. Weis W, Brown JH, Cusak S, Paulson JC, Skehel JJ, Wiley DC: Structure of the influenza virus hemagglutinin complexed with its receptor, sialic acid. *Nature* 1988; 333:426.
37. Williams WV, Kieber-Emmons T, Weiner DB, Greene MI: Contact residues and predicted structure of the reovirus type 3-receptor interaction. Presented at "Modern Approaches to New Vaccines Including the Prevention of Aids", September 14–18, Cold Spring Harbor Laboratory, NY, 1988, Abstract #123.
38. Gallo RC, Sarin PS, Gelmann EP, et al: Frequent detection and isolation of cytopathic retroviruses from patients with AIDS and at risk for AIDS. *Science* 1983;220:865.
39. Barré-Sinoussi F, Chermann JC, Rey F, et al: Isolation of a T lymphotrophic retrovirus from a patient at risk for acquired immune deficiency syndrome (AIDS). *Science* 1989; 220:868.
40. Lane H, Fauci A. Infectivity and Pathogeneisis of AIDS. *Ann Rev Immunol* 1985; 3:477.
41. Kowalski M, Potz J, Basiripour L, et al: Functional regions of the envelope glycoprotein of human immunodeficiency virus type 1. *Science* 233; 209–212.
42. Goulsmit J: Immunodominant B-cell epitopes of the HIV-1 envelope recognized by infected and immunized hosts. *AIDS* 1988; 2s:41.
43. Klatzmann D, Champagne E, Chamaret S, et al: T lymphocyte T-4 behaves as the receptor for the human retrovirus LAV. *Nature* 1984; 312:767.
44. Dalgleish AG, Beverley PC, Clapham PR, Crawford DH, Greaves MF, Weiss RA: The CD4 (T4) antigen is an essential component of the receptor for the AIDS retrovirus. *Nature* 1984; 312:763.
45. Maddon RJ, Dalgleish AG, McDougal JS, Clapham PR, Weiss RA, Axel R: The T4 gene encodes the AIDS virus receptor and is expressed in the immune system and the brain. *Cell* 1985; 47:333.
46. Sattentau QJ, Dalgleish AG, Weiss RA, Beverley PCL: Epitopes of the CD4 antigen and HIV infection. *Science* 1986; 234:1120.
47. Jameson BA, Rao PE, Kong LI, et al: Location and chemical synthesis of a binding site for HIV-1 on the CD4 protein. *Science* 1988; 240:1335.

48. Landau NR, Warton M, Littman DR: The envelope glycoprotein of the human immunodeficiency virus binds to the immunoglobulin-like domain of CD4. *Nature* 1988; 334:159.
49. Peterson A, Seed B: Genetic analysis of monoclonal antibody and HIV binding sites on the human lymphocyte antigen CD4. *Cell* 1988; 54:65.
50. Lasky LA, Nakamura G, Smith DH et al: Delineation of a region of the human immunodeficiency virus type 1 gp120 glycoprotein critical for interaction with the CD4 receptor. *Cell* 1987; 50:975.
51. Weiner DB, Williams WV, Hoxie JA, Berzofsky JA, Greene MI: Non CD4 molecules on human cells Important in HIV-1 human-cell interactions. *Vaccines 89*. New York, Cold Spring Harbor Press, 1989, p 114.
52. Weiner DB, Hubner K, Williams WV, Greene MI: Species trophism of HIV-1: Infectivity of interspecific cell hybridomas implies non-CD4 structures are required for cell entry. *Cancer Detect Prev* 1990; 14:317–320.
53. Clapham PR, Weber JN, Whitby D, et al: Soluble CD4 blocks HIV-1 infection of T cells and macrophages but not fibroblast cell lines. *Nature* 1989; 337:368.
54. Weiner DB, Williams WV, Hoxie JA, Greene MI: Identification of ancillary molecules on human T cells important to HIV-1 T cell interactions. IV International Conference on AIDS, Stockholm, June 1988, abstract 2577.
55. Weiner DB, Williams WV, Merva MJ, Huebner FK, Berzofsky JA, Greene MI: HIV infectivity: Analysis of virus envelope determinants and target cell requirements for infectivity by HIV-1. *Vaccines 90*. New York, Cold Spring Harbor Press, 1990, p. 339–345.

CHAPTER 9

Anti-Idiotypic Antibodies: An Alternative Approach to Immunoprophylaxis Against Parasites

Jean-Marie Grzych, Florence Velge-Roussel, and André Capron

Introduction

Parasitic infections represent one of the major worldwide public health problems. These infections, which are essentially confined to the developing countries, affect more than one thousand million people and are directly or indirectly involved in about one million deaths each year. In addition to these dramatic humanitarian aspects, parasitic infections also profoundly affect the economic development of countries in endemic regions.

It is therefore not surprising that the elaboration of safe and efficient vaccines against parasitic infections has been the primary goal of various national and international programs since the early days of immunological research in parasitology.

During the last decade, important progress has been made in our understanding of parasite immunity. These studies have allowed the identification of the main cellular and/or humoral components involved in immunity to parasites. It is now becoming clear that host-parasite relationships are finely regulated, on the one hand, by the ability of the host to mount an effective immune response to eliminate the parasite, and on the other hand, by the mechanisms developed by the parasite in order to circumvent the host effector functions and to allow its survival.

In the context of this reciprocal selective pressure, the administration of well-defined parasite antigens employed as vaccines may disrupt the equilibrium in favor of the host. Along these lines the close relation established between the expression of particular cellular and/or humoral effector mechanisms and the acquisition of a partial or total immunity in the corresponding host makes the target antigens of these mechanisms potential candidates for parasite immunoprophylaxis.

Up until now, two essential factors have limited this approach: the characterization and the isolation of these target antigens and their large-scale production. In the case of protozoan or helminth parasites, these molecules are found in different developmental stages, such as intermediate host stages or intracellular forms, which cannot be obtained in sufficient quantities from natural sources. Nevertheless, this area has moved forward considerably since the introduction into immunoparasitology of hybridoma technology, which allows the production of monoclonal antibodies with well-defined biological functions, and the development of molecular biology for the production of recombinant proteins.

Despite the real efficiency of recombinant DNA methodology, such techniques remain inapplicable in the case of carbohydrate or lipid antigens. In this respect, the application of the fundamental concept of Jerne's (1) theory, which proposes that certain anti-idiotype (anti-Id) antibodies bear the internal image of the antigen, offers an alternative approach to study the immunological properties of non-peptidic epitopes involved in antiparasite immunity. In this chapter, we will illustrate this topic with some of the encouraging results acquired with monoclonal or polyclonal anti-Id preparations specific for monoclonal antibodies (MAb) that exhibit marked in vitro and/or in vivo effector functions in parasites. We shall also discuss the possibilities offered by anti-Id antibodies in the fine dissection of immunity to parasites.

Anti-idiotypic Immunization in Protozoan Infections

Immunization Against African Trypanosomes

Trypanosomiasis was the first parasitic disease for which the potential applicability of anti-Id immunizations was clearly demonstrated (2). In this model, anti-Id antibodies have been raised against those protective MAb specific for *Trypanosoma rhodesiense* variable antigens (VAT) that were capable of neutralizing parasites in vivo. It was shown that injection of a purified IgG1 fraction of these polyclonal anti-Id antibodies collected from SLJ mice, was able to modify the course of primary parasitemia in an animal model.

The immunity induced was manifested either by a reduction of the first peak of parasitemia or by a selection against parasites expressing the corresponding VAT. The injection of anti-Id antibodies was associated with the generation of antigen-specific antibodies that shared idiotypic cross-reactivity with one of the initial MAb (7H11-Id).

Further studies (3), in this model, indicate that the induction of immunity is restricted to mice bearing genes linked to IgH_c^a. Only such mice respond to immunization with anti-Id antibodies and show enhanced expression of the 7H11-Id (3). These latest data support the fact that these antibodies are unlikely to be internal images of antigens because they fail to bind conventional rabbit antisera specific for the initial antigen but also because the anti-7H11 antibodies fail to induce immunity or idiotype expression in all strains of mice. Under these conditions, the anti-Id antibodies produced seem to recognize individual idiotopes (IdI) but not cross-reactive idiotopes (IdX).

Even though they seem to be specific for paratope-associated idiotopes, they fail to induce immunity across genetic barriers.

Molecular Mimicry of Carbohydrate Epitopes

More recently, Sacks et al (4) have extended their observations to an intracellular trypanosome: *Trypanosoma cruzi*, the etiological agent of Chagas' disease. In this model, rabbit polyclonal anti-idiotype antibodies have been prepared against the WIC29.26 MAb specific for an unusual carbohydrate epitope of the gp72 glycoprotein of *T. cruzi* (5). This antigen (MW 72 kDa) presents on the surface of *T. cruzi* epimastigotes, and metacyclic trypomastigotes (5) is able to protect mice against an insect-derived challenge with the parasite (6).

Immunization of BALB/c mice with the affinity-purified rabbit polyclonal anti-Id antibodies induced high levels of anti-gp72 antibodies detectable either by indirect immunofluorescence assays on the 29.26 *T. cruzi* positive strain or by immunoprecipitation of [125I]-labeled parasites. However, these antibodies fail to react with *T. cruzi* strains that express the gp72 antigen but do not expose the 29.26 epitope (7), which therefore shows that these antibodies are specific for the carbohydrate determinant of gp72.

More interestingly, immunization of rabbits or guinea pigs with this anti-Id polyclonal gives a strong anti-gp72 antibody response. These data suggest either that the 29.26-Id represents an interspecies cross-reactive idiotope (IdX), or that the anti-Id preparation contains internal images of the initial carbohydrate epitope and acts as a surrogate antigen. The latter alternative appears to be more likely in the context of further observations in which the binding of the ^{125}I-labeled 29.26 MAb to the anti-Id antibodies was shown to be inhibited by the addition of rabbit anti-gp72 antibodies or by *T. cruzi*-infected human sera.

Anti-Id Antibodies and Cell-Mediated Immunity

Several protozoan species that are major pathogens in man have a predominantly intracellular location. Such organisms are shielded from the action of specific antibodies and their elimination depends on the recognition of the parasite antigens on the host cell surface, particularly by specific T cells. It is therefore essential that vaccines developed against such infections contain epitopes involved in T cell recognition.

The efficiency of anti-Id antibody preparations has also been improved in this domain, since several authors have reported that B and T cells share idiotypic determinants (8, 9). These observations, initially established in viral and bacterial infections, have been amply confirmed in the parasite model murine *Leishmania major* infection (10). *Leishmania major* induces in man a spectrum of diseases ranging from self-healing cutaneous lesions to fatal visceral infections. These parasites exclusively reside in macrophages and require cell-mediated immunity for their control.

In this model, a potential candidate for immunoprophylaxis has been identified as a glycolipid antigen present on the membrane of the insect stage parasite and also on the surface of parasitized macrophages (11). This molecule plays a major role in the initial attachment of the parasite to the macrophage and is directly involved in the initiation of infection (11). This purified glycolipid is able to protect genetically susceptible mice against a fatal infection with *L. major* (12).

An anti-Id polyclonal preparation specific for the 79.3 MAb directed against this glycolipid protects susceptible mice against a challenge infection with *L. major*. Interestingly, the level of antibodies specific for the parasite detected after anti-Id immunization is very low and suggests that these mice were able to control their infection through cellular mechanisms primed by anti-Id immunization.

The use of anti-Id antibodies to prime T cells has been previously reported for *Listeria monocytogenes* (13) and in Sendai virus infection (14). In both cases anti-Id antibodies preferentially stimulate T cell epitopes rather than B cell epitopes. If from these data it is clear that anti-Id immunization can stimulate protective cell-mediated immunity, it is not yet known how the anti-Id structures interact with class II molecules for T cell recognition, especially in the case of internal images of nonpeptidic epitopes.

Anti-Id Immunization in *Schistosoma mansoni* Infection

Schistosomiasis Acquired Immunity and ADCC Mechanisms

Anti-Id immunizations have been also successfully applied in our laboratory for *Schistosoma mansoni* immunoprophylaxis. These studies have allowed us to investigate further two major *S. mansoni* antigens previously identified as potential protective immunogens but inaccessible to recombinant DNA technology.

The development of an anti-Id strategy was closely dependent on the particular context of *S. mansoni* immunity in which the young schistosome (the schistosomulum) was clearly defined as the main target of several effector mechanisms of acquired immunity. These ADCC (antibody-dependent cellular cytotoxicity) mechanisms were demonstrated in the rat model but are also present in monkeys and in man and are the major effector mechanisms of acquired immunity (15). Three main cell populations (macrophages, eosinophils, and platelets) have been shown to kill the schistosomulum in cooperation with anaphylactic antibodies (IgE and IgG subclasses), previously demonstrated to be the major humoral factors of acquired immunity (16).

As we mentioned above, the close correlation between the ADCC mechanisms and the expression of anti-schistosome immunity led us to produce MAb with well-defined effector functions and to use them to characterize and to isolate the target antigens of these mechanisms.

Anti-*S. mansoni* MAb have been produced in a rat × rat homologous system of hybridization using the IR983F myeloma cell line (17) and the splenic cells from LOU rats either infected with *S. mansoni* or immunized with different antigenic fractions of the parasite.

Induction of In Vitro and In Vivo Protective Ab3 Responses

In an initial series of hybridizations performed with splenic cells of LOU rats experimentally infected with *S. mansoni* cercariae, we isolated one hybridoma that secreted a specific anti-*S. mansoni* antibody, IPL Sm1. This antibody of the IgG2a isotype exhibits a high level of cytotoxicity in the presence of normal rat eosinophils and confers a significant degree of protection when passively transferred to naive rats (18). The IPL Sm1 was shown to bind specifically to a 38-kDa schistosomulum surface antigen (19) defined as a major immunogen in schistosome infections of various animal species including man (20).

Although the 38-kDa antigen theoretically is a good candidate for a vaccine against schistosomiasis, this antigen induces the production of specific antibodies in 97% of Brazilian *S. mansoni*-infected patients (21), since the glycanic nature of the epitope recognized by the IPL Sm1 antibody precludes its production by recombinant DNA methodology.

Moreover, it was shown that, together with protective antibodies, the 38-kDa molecule was able to induce the production of blocking IgG2c antibodies that inhibit, in vitro and in vivo, the functional properties of the IPL Sm1 antibody (22). Therefore, on the basis of the Jerne theory, we considered as an alternative approach the production of anti-Id antibodies against the IPL Sm1 antibody.

Anti-Id MAb to IPL Sm1 were obtained by the fusion of the IR983F rat myeloma cell line and spleen cells from LOU rats previously immunized with purified IPL Sm1 antibody (23). From 200 hybrid cell supernatants obtained in two successive cell fusion experiments, 29 supernatants inducing significant levels of inhibition (> 70%) were selected. The inhibitory activity of cell supernatants on IPL Sm1 binding to the 38-kDa antigen strongly indicated that the MAb produced were able to bind to an epitope of the IPL Sm1 antibody that was close to a part of the combining site. This further suggested that Ab2 antibodies could be paratope-induced antibodies and might bear an internal image of the original epitope.

Immunization experiments carried out with one particular clone (JM8-36) producing Ab2 antibody of the IgM isotype (23) reproduced several parameters of *S. mansoni* infection. Successive injections of the JM8-36 antibody elicited anti-*S. mansoni* Ab3 antibodies in the naive LOU rat, which were capable of inhibiting the binding of [^{125}I] labeled IPL Sm1 antibody to its target antigen. This indicated specificity for the same molecule (Fig. 9.1). These observations were confirmed by the demonstration that Ab3 antibodies

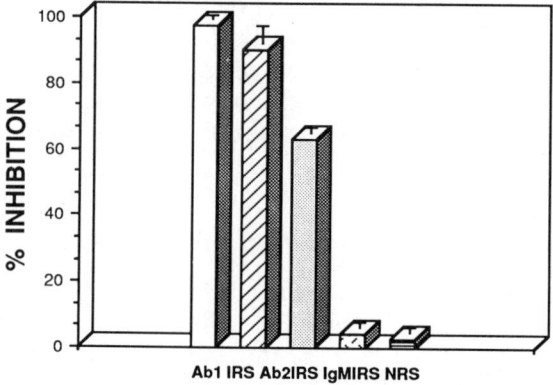

Fig. 9.1. Inhibition of Ab1 antigen-binding by Ab3 antibodies. Inhibition assays of radiolabeled Ab1 antibody (IPLSm1) to the 38-kDa antigen were performed (23). 50 μl of ^{125}I-labeled IPL Sm1 were incubated with 50 μl of inhibiting factor: IPL Sm1 antibody (Ab1), four-week-infected rat serum (IRS), Ab2-immunized rat sera (Ab2IRS), unrelated IgM immunized rat sera (IgM IRS), or normal rat serum (NRS) (mean of two duplicate experiments ± SD).

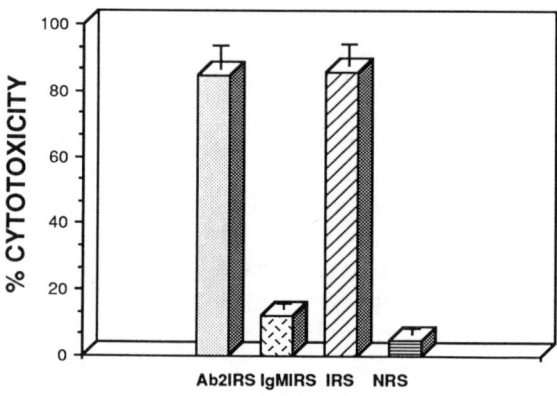

Fig. 9.2. In vitro effector function of polyclonal Ab3. Eosinophil-dependent cytotoxicity experiments were performed (24) on the sera of rats previously immunized with JM8-36 MAb (AB2IRS) or sera from rats immunized with unrelated IgM antibodies C (IgM IRS). Eosinophil-dependent cytotoxicity was measured after a 48-hour contact of skin schistosomula and preincubated overnight with the sera at a final dilution of 1/16 with a rat eosinophil-rich population (40 to 60% eosinophils). The percentage of cytotoxicity was compared at an equivalent dilution with control sera, normal rat serum (NRS), or serum from four-week-*S. mansoni*-infected rats (IRS) (mean of two duplicate experiments ± SD).

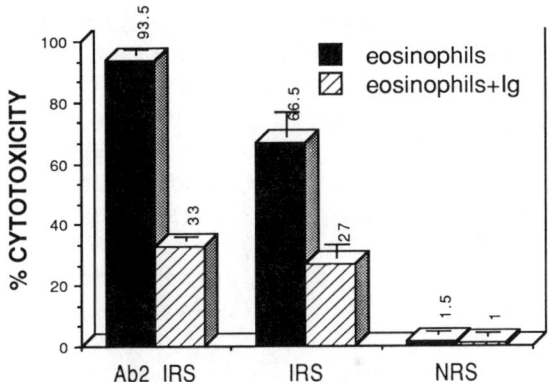

Fig. 9.3. Involvement of IgG2a antibodies in Ab3-mediated cytotoxicity. Skin schistosomula were incubated overnight with JM8-36 immunized rat sera (Ab2IRS) (1/16 final dilution). The percentages of cytotoxicity were evaluated after a 48-hour incubation of schistosomula in the presence of a rat eosinophil-rich cell population incubated 2 hour either with MEM medium of IgG2a aggregated rat myeloma proteins. The percentages of cytotoxicity were compared to controls (four-week-infected rat sera, IRS) and normal rat sera (NRS) studied under the same conditions (mean of two duplicate experiments ± SD).

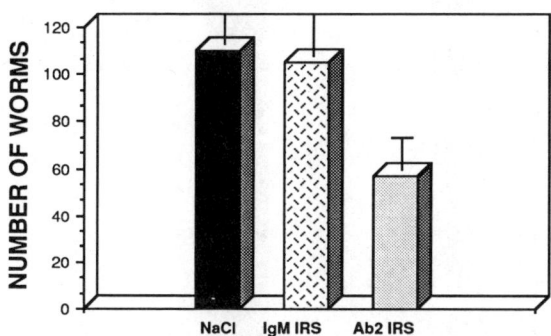

Fig. 9.4. Passive transfer of Ab3 antibodies. One milliliter of serum collected from JM8-36 immunized rats (Ab2IRS) or from rats immunized with normal IgM (IgMIRS) was injected intravenously into LOU rats infected 4 hours previously with 800 S. mansoni cercariae. Parasite burdens were evaluated three weeks later by the liver perfusion technique. The numbers of worms obtained from these groups were compared with the parasite burden of rats injected with 1 ml of physiological saline (NaCl). The percentage of protection was calculated by the formulae $(a - b)/a \times 100$, where a is the number of worms recovered from the saline-injected control group and b is the number of worms recovered from a rat injected with 1 ml of Ab3 serum (Ab2IRS).

obtained by immunization with Ab2 elicited anti-*S. mansoni* antibodies that could be detected by indirect immunofluorescence reactions on cryopreserved schistosomulum sections.

The Ab3 antibodies produced by immunization with Ab2 induced a significant level of cytotoxicity (70–90%) in the presence of normal rat eosinophils. This range of killing was observed using either 4-week-infected rat sera or the IPL Sm1 MAb and reproduced the ADCC mechanism previously described by Capron et al (24) as one of the essential effector systems of immunity in rat infection (Fig. 9.2). More recently, we have shown that the eosinophil-dependent cytotoxicity mediated by these Ab3 antibodies could be significantly decreased when effector cells, that it, eosinophils, were preincubated in the presence of aggregated IgG2a myeloma protein, which suggests that the cytotoxic Ab3 antibodies were of the IgG2a isotype (Fig. 9.3).

The biological relevance of these in vitro findings was amply confirmed in vivo with the observation that the passive transfer of polyclonal Ab3 antibodies decreased the parasitic burden of recipient animals (Fig. 9.4). Most importantly, it was established that naive LOU rats immunized with Ab2 demonstrated a marked protection against a subsequent cercarial exposure. This clearly suggests the potential use of anti-idiotype antibodies to enhance a possible vaccination against schistosomes (Fig. 9.5).

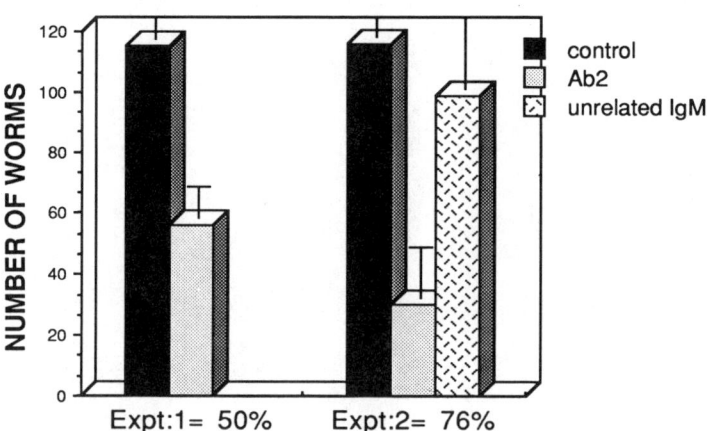

Fig. 9.5. Protective effect of Ab2 immunization. Male LOU rats immunized with JM8-36 antibody were infected with 800 *S. mansoni* cercariae. The parasite burdens were measured three weeks later by liver perfusion. The number of worms obtained from the rats immunized with JM8-36 antibody (Ab2) were compared to those obtained in the control group (LOU rats injected with physiological saline—control—or with IgM purified from NRS—unrelated IgM). The percentage of protection was calculated by the formula $(a - b)/a \times 100$, where a is the number of worms obtained from the saline-injected control and b is the number of worms recovered from Ab2-immunized rats.

In addition to this first set of experimental data, the existence of cross-reactive idiotopes between rat and human models was investigated. The immunofluorescence test, previously described by Thanavala et al (25), and applied in this model, revealed that antibodies present in the sera of *S. mansoni*-infected patients specifically bind to the Ab2 antibodies present on the membrane of JM8-36 hybridoma cells. Such observations were confirmed in a radioimmunoassay by the inhibition of ^{125}I-labeled IPL Sm1 binding to the JM8-36 Ab2 antibody by the sera of *S. mansoni*-infected patients, which indicates the presence in these sera of antibodies capable of binding the Ab2 antibody specifically.

These results, together with the fact that similar effector mechanisms have been described in both the rat and in man, suggest that one can consider favorably the use of anti-idiotypes in man to elicit anti-*S. mansoni* antibody responses.

Isotypic Selection and Idiotypic Immunization

Although the experimenal data reported above for the JM8-36 monoclonal Ab2 antibody provided new evidence for the efficiency of anti-idiotype immunization in parasite models, they also raised several questions about the biological significance of the idiotypic cascade during the time-course of parasitic infection.

The fact that the injection of an anti-idiotype antibody (Ab2) specific for a rat monoclonal IgG2a antibody induces anti–anti-Id antibodies (Ab3) expressing the same functional properties as the initial Ab1 suggests a possible correlation via the idiotypic cascade between isotypic selection of Ab3 and the nature of the initial Ab1 idiotopes and/or isotypes.

In order to answer these questions, we recently undertook a set of experiments to define the role of Ab2 and Ab3 antibodies in experimental rat schistosomiasis. These antibodies were raised after the injection into syngeneic rats of an IgE MAb (B48-14) (26). We placed particular emphasis on isotypic selection and consequently on the effector functions of the Ab3 antibodies.

The B48-14 MAb was obtained by the fusion of splenic lymphocytes from LOU rats immunized with a glycoprotein-enriched fraction of *S. mansoni* incubation products and the IR983F myeloma cell line. This IgE antibody, when tested in effector cell-mediated cytotoxicity, revealed a strong capacity to kill schistosomula in the presence of normal rat macrophages, eosinophils, and platelets, comparable to the levels of mortality routinely obtained with infected rat sera. As in the case of the IPL Sm1 antibody, B48-14 antibody passively transferred to normal recipient rats was able to provide a significant level of protection, ranging between 40 and 50%, against a challenge infection with *S. mansoni* cercariae.

Interestingly, the B48-14 IgE antibody recognizes the 26-kDa larval antigen present in the schistosomula-release products antigens (SRP-A). This material, which is excreted or secreted by schistosomula, was previously reported to be directly involved in the production of cytotoxic and protective polyclonal IgE antibodies against three main antigens, of 56, 26, and 22 kDa. These molecules also supported the protection conferred to normal rats either by passive transfer of polyclonal IgE or after immunization with SRP-A (27, 28).

Immunization of LOU rats with B48-14 antibody has been shown to induce the production of specific anti-idiotype antibodies (Ab2) capable of binding to the Ab1 antibody. Time-course studies of these immunizations allowed us to demonstrate that the first wave of Ab2 antibodies is followed by the appearance of antibodies exhibiting a clear anti-*S. mansoni* activity, which suggests therefore that the anti-idiotype antibodies (Ab2) elicit the production of "natural" anti-anti-idiotype antibodies (Ab3) specific for the parasite (29).

Further analysis of the Ab3 response using specific antisera to rat immunoglobulins revealed the existence of two successive peaks of Ab3 antibodies. The first was mainly IgG followed three weeks later by an IgE Ab3 antibody peak. In the latter case, the levels of anti-schistosomula, IgE-mediated cytotoxicity in the presence of normal rat platelets was comparable to that observed with a 42-day *S. mansoni* infection serum (30) (Fig. 9.6).

The investigation of Ab3 antibody specificity by immunoprecipitation and Western blotting analysis confirmed the antiparasite activity of these Ab3 antibodies. The IgG Ab3 strongly precipitate the 56-kDa antigen and the IgE Ab3 preferentially recognizes the 26-kDa molecule in SRP-A preparations, which suggests the existence of natural β Ab3 in the sera tested (29). This

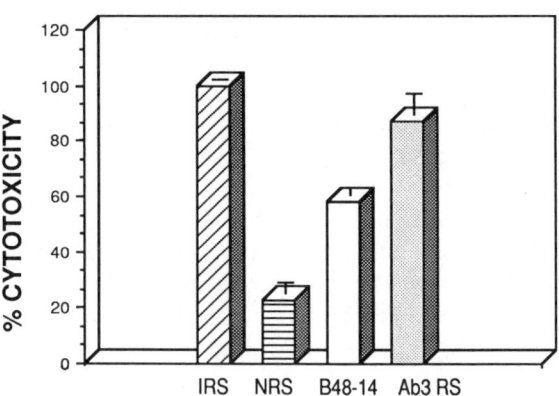

Fig. 9.6. Effector functions of polyclonal IgE Ab3 antibodies on platelet-mediated cytotoxicity. Platelet-dependent cytotoxicity assays were performed (22). Platelets from normal Fischer rats were incubated with 50 schistosomula in the presence of sera; either 20 μl of infected rat sera (IRS), 20 μl of normal rat sera (NRS), or 20 μl of the IgE Ab3 peak (Ab3IRS). For the B14-48Ab1 antibody, a cell culture supernatant of the hybridoma was used in the presence of 10% normal rat serum. The percentage of cytotoxicity was measured after a 24-hour incubation in 5% CO_2 at 37°C, by microscopic examination.

hypothesis is also supported by the close parallelism observed between the level of IgE Ab3 antibodies specific for the 26-kDa antigen and the expression by the same sera of a marked platelet-dependent cytotoxicity, which reproduces the killing activity of B48-14 monoclonal IgE antibody (Fig. 9.6). Moreover, Ab1- and Ab3 IgE-dependent platelet cytotoxicity could be inhibited by the addition of polyclonal Ab2 (Fig. 9.7). These data support the view that Ab1 and Ab3 recognize the same epitope on the 26-kDa schistosomulum antigen.

The immunological and functional identity between B48-14 Ab1 and polyclonal Ab3 IgE was further illustrated in vivo, since passive transfer of Ab3 IgE antibodies was shown to reduce (by 50%) the parasite burden of naive LOU rats challenged by *S. mansoni* cercariae (Fig. 9.8).

Together with a further demonstration of the possibilities offered by anti-idiotype antibodies in *S. mansoni* immunoprophylaxis, these results provide essential information concerning the isotypic pattern of Ab3 antibodies produced subsequent to an Ab1 immunization. The main observation in this field is the isolation of IgE Ab3 antibodies that display marked anti-*S. mansoni* effector functions comparable to those previously reported for the B48-14 MAb (Ab1) (26).

Such observations raise questions concerning the isotype selection of antibodies through the idiotypic cascade, and one can ask whether the Ab3

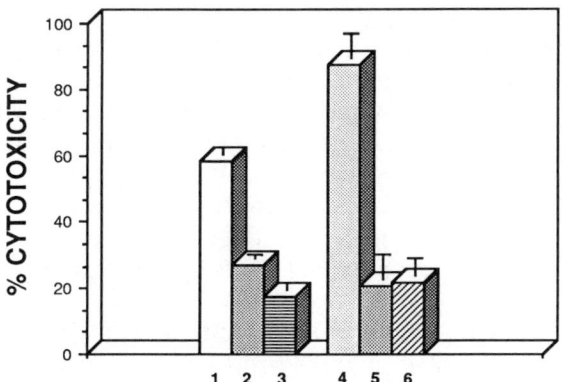

Fig. 9.7. Inhibitory effect of Ab2 antibodies on platelet-mediated cytotoxicity. The antagonistic effect of Ab2 antibodies on the effector functions of Ab1 and Ab3 antibodies was studied in the presence of normal rat platelets previously incubated for 30 minutes at 37 °C in the presence of heated Ab2 rat sera (column 3 Ab1 + Ab2, column 6 Ab3 + Ab2). Then Ab1 (B48-14 cell supernatant) or Ab3 rat sera were added together with the schistosomula. The percentage of cytotoxicity was measured after a 24-hour incubation period. Columns 1 and 4 correspond, respectively, to the cytotoxic activity of Ab1 and Ab3 in the presence of platelets pretreated with heated infected rat sera. Columns 2 and 5 rpresent the cytotoxic activity of Ab2 antibodies alone (mean of two duplicate experiments ± SD).

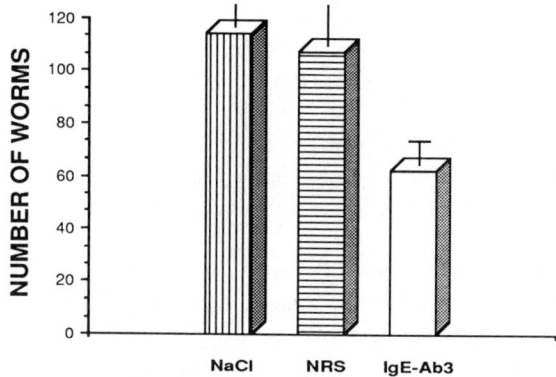

Fig. 9.8. In vivo protective role of IgE polyclonal Ab3 antibodies. One microliter of Ab1-immunized rat sera (Ab3 peak) (IgE-Ab3) was injected intraperitoneally into male LOU/M rats subsequently infected with 1000 *S. mansoni* cercariae. Parasite worm burdens were evaluated 21 days after the challenge infection by liver perfusion. Control animals were injected with 1 ml physiological saline solution (NaCl) or normal rat serum (NRS) (1 ml/rat). The percentage of protection was calculated by the formula $(a - b/a) \times 100$, where a is the number of worms recovered from rats injected with physiological saline solution and b is the number of worms collected from rats injected with IgE Ab3 rat sera (mean of two experiments).

isotype or idiotype could be related to the isotype and/or idiotype nature of the initial Ab1 used for immunization. In other words, does the synthesis of Ab3 antibodies with anti-*S. mansoni* activity depend on the structure of the epitope recognized on the 26-kDa antigen by the Ab1 antibody, an epitope also mimicked on the Ab2 β antibodies? (According to this hypothesis, such an epitope would orientate the Ab3 response towards antibodies sharing the same specificity and isotype as the Ab1). Alternatively, does it mean that the isotypic selection of Ab3 antibodies is directly influenced by the Fc_E portion of the Ab1 antibody? This latter point could be answered by immunization experiments involving the injection of purified Fab'2 fragments of the B48-14 MAb.

However our experimental data show that independently of the low theoretical level of Ab3 β antibodies, we could observe an efficient and conserved biological activity along the idiotypic cascade. It seems that the dilution of antibodies during clonal expansion might be corrected by the selection of anti-idiotype antibodies exhibiting strong biological functions. The anti-Id network should be restricted in antibody response toward epitope/paratope couples of high affinity and immunological efficiency (31).

Moreover, the anti-Id network with its capacity for continuous endogeneous stimulation may also provide a further mechanism for long-term memory (31), which in the case of a vaccine strategy might have profound practical consequences in the development of long-term protective immunity toward schistosomes.

Finally, these results confirmed that the 26-kDa molecule and particularly the epitope recognized by the B48-14 MAb have immunoprophylactic potential.

Conclusion

At this stage of research, it seems difficult to anticipate the applicability of an anti-idiotype strategy in the production of safe protective preparations for human immunoprophylaxis. Indeed, the use of anti-idiotypes as surrogate vaccines in humans is obviously hampered by several factors. First of all, the injection of heterologous anti-idiotype antibodies in man could be considered unethical. These foreign proteins could elicit an unexpected immune response such as serum sickness, rapid clearance of anti-idiotype antibodies, or induction of immune diseases at the result of an idiotype–anti-idiotype interaction. Another limitation of the anti-Id antibodies is the selection of certain idiotypes and epitopes as unique targets, which could involve the risk of generating suppression or inefficiency resulting from genetic selection. Additionally, one cannot exclude the possibility that some of the injected Anti-idiotype antibodies would react as anti-cell receptor antibodies, particularly in the case of anti-idiotype antibodies raised against Ab1, specific for epitopes involved in the binding to host cell membrane receptors.

Some of these side effects could be limited if chimeric antibodies using rat or mouse VH domains and the CH domains of human immunoglobulins were constructed (32). Other possibilities include the production of human Anti-idiotype antibodies or the isolation, after appropriate treatment, of the polypeptide structure that bears the internal image of the initial epitope.

In spite of the real progress made in experimental infections, many questions need to be answered before anti-idiotype methodology could be applied in humans. It is necessary to define the dose and posology of anti-idiotype injections to obtain a strong, long-costing immunity. This particular problem of immunizing doses is not yet resolved because of the difficulty in determining the appropriate amount of anti-Id antibodies to be injected. This is closely dependent on anti-Id isotypes, the degree of mimicry between the anti-Id antibodies and the initial epitope, and the ability of anti-Id antibody to bind to the immunocompetent cells. All these aspects need now to be studied in various experimental models, including baboons, before any attempt at immunization in man.

However, the use of anti-idiotype preparations can also offer some advantages as compared to conventional approaches to the development of vaccines.

Anti-Id antibodies are less hazardous than attenuated vaccines because they do not carry the risk of conversion to pathogenicity.

The anti-Id network provides a new opportunity to obtain peptidic internal images of nonpeptidic epitopes for which production by recombinant DNA methodology is at present impossible. Those antibodies that mimic carbohydrate epitopes may overcome unresponsiveness to the nominal antigen and limit the delay in response towards glycanic antigens that is observed in young children. An earlier study (33) has already demonstrated the feasibility of this approach: The injection of anti-Id antibodies prepared against a specific, MAb for a bacterial levan induced a strong IgM and subsequent IgG response in CBA/N mice bearing an Id X deficiency when treated at birth with anti-Id antibodies.

Moreover, the demonstration that anti-Id antibodies can activate T cells is particularly encouraging, especially in the case of intracellular pathogens in which cell-mediated immunity is required. Also synthetic or recombinant antigens may be less able to induce these effects, presumably due to differences in antigen processing and presentation or because they lack intrinsic T cell sites.

Together with the potential application of anti-Id antibodies in parasite immunoprophylaxis, it is appropriate, before concluding this chapter, to mention several domains in which anti-Id antibodies might represent powerful tools for the enhancement of our understanding of parasitic infections.

The initial report of Potocnjak et al (34), in which an anti-Id antibody specific for a protective MAb directed against the Pb44 protein of *Plasmodium berghei* was successfully used in the characterization of an epitope in a crude extract of the mosquito salivary gland, makes anti-Id antibodies a new tool for the detection of circulating antigens in biological fluids.

Anti-Id antibodies also offer new ways to approach the fine dissection of the anti-parasitic immune response. Indeed the opportunity offered through the internal image concept to follow the development of the antibody isotype response against a particular epitope is one example. These investigations now appear to be essential, since the demonstration in rat and human schistosomiasis of particular isotypes that exhibit blocking or modulatory activities (22–35). Finally, anti-Id antibodies would certainly contribute greatly to the study of the possible relation between the particular conformational structure of an epitope and the preferential induction of a particular antibody isotype.

Acknowledgments. This work was supported by INSERM U 167 and CNRS 624, and by grant n°07585 from the Edna McConnel Clark Foundation. We wish to thank Dr. RJ Pierce for his critical reading of the manuscript.

References

1. Jerne NK: Towards a network theory of the immune system. *Ann Immunol Inst Pasteur* 1974; 125C:373.
2. Sacks DL, Esser KM, Sher A: Immunization of mice against African trypanosomiasis using anti-idiotypic antibodies. *J Exp Med* 1982; 155:1108.

3. Sacks DL, Sher A: Evidence that anti-idiotype induce immunity to African trypanosomiasis is genetically restricted and requires recognition of combining site related idiotypes. *J Immunol* 1983; 131:1511.
4. Sacks DL, Kirchoff LV, Hieny S, Sher A: Molecular mimicry of carbohydrate epitope on a major surface glycoprotein of *Trypanosoma cruzi* using anti-idiotypic antibodies. *J Immunol* 1985; 135:4155.
5. Snary D, Fergusson MAJ, Scott P, Allan AK: Cell surface antigens of *Trypanosoma cruzi*: Use of monoclonal antibodies to identify and isolate an epimastigote specific glycoprotein. *Mol Biochem Parasitol* 1981; 3:343.
6. Snary D: Surface glycoproteins of *Trypanosoma cruzi*: Protective immunity in mice and antibody levels in chagasic sera. *Tr R Soc Trop Med Hyg* 1983; 77:126.
7. Kirchoff LV, Hieny S, Shriver GM, Snary D, Sher A: Cryptic epitope explains the failure of monoclonal antibody to bind to certain isolates of *Trypanosoma cruzi*. *J Immunol* 1984; 133:2731.
8. Eichman K, Ben-Neriah Y, Hetzerlberger D, Polke C, Givol D, Lonai P: Correlated expression of Vh framework and Vh idiotypic determinants on T helper cells and functionally undefined T cells binding group A streptococcal carbohydrate. *Eur J Immunol* 1980; 10:105.
9. Woodland R, Cantor H: Idiotype-specific T helper cells are required to induce idiotype-positive B memory cells to secrete antibody. *Eur J Immunol* 1978; 8:600.
10. Sacks DL: Immunization against parasitic protozoa using anti-idiotypic antibodies. *Monogr Allergy* 1987; 22:166.
11. Handman E, Goding JW: The *Leishmania* receptor for the macrophage is a lipid-containing glycoconjugate. *Eur Mol Biol Org J* 1985; 4:329.
12. Handman E, Mitchell GF: Immunization with *Leishmania* receptor for macrophages protects mice against cutaneous leishmaniasis. *Proc Natl Acad Sci USA* 1985; 82:5910.
13. Kaufmann SHE, Finberg RW, Miller I, Wrazel LJ: Vaccination against *Listeria monocytogenes* with clonotypic antiserum. *J Immunol* 1985; 134:4123.
14. Ertl HCJ, Finberg RW: Sendai virus specific T cell clones: Induction of cytolytic T cells by an anti-idiotypic antibody directed against a helper T cell clone. *Proc Natl Acad Sci USA* 1984; 81:2850.
15. Capron M, Capron A: Rats, mice and men models for immune effector mechanisms against schistosomiasis. *Parasitol Today* 1986; 2:69.
16. Capron A, Dessaint JP: Effector and regulatory mechanisms in immunity to schistosomes: A heuristic view. *Ann Rev Immunol* 1985; 3:455.
17. Bazin H, Grzych JM, Verwaerde C, Capron A: A LOU rat non-secreting myeloma cell line suitable for the production of rat-rat hybridomas. *Ann Immunol* 1980; 131D:359.
18. Grzych JM, Capron M, Bazin H, Capron A: In vitro and in vivo effector functions of rat IgG2a monoclonal anti-*S. mansoni* antibodies. *J Immunol* 1982; 129:2739.
19. Dissous C, Grzych JM, Capron A: *Schistosoma mansoni* surface antigen defined by a rat protective monoclonal IgG2a. *J Immunol* 1982; 129:2232.
20. Dissous C, Capron A: Isolation of surface antigens from *Schistosoma mansoni* Schistosomula. In: Peeters H, ed: *Protides of Biological Fluids*. Oxford and New York: Pergamon Press; 1982; pp 179.
21. Dissous C, Prata A, Capron A: Human antibody response to *Schistosoma mansoni* surface antigens defined by protective monoclonal antibodies. *J Infect Dis* 1984; 149:227.

22. Grzych JM, Capron M, Dissous C, Capron A: Blocking activity of rat monoclonal antibodies in experimental schistosomiasis. *J Immunol* 1984; 133:998.
23. Grzych JM, Capron M, Lambert PH, Dissous C, Torres S, Capron A: An anti-idiotope vaccine against schistosomiasis. *Nature* 1985; 316:74.
24. Capron M, Capron A, Torpier G, Bazin H, Bout D, Joseph M: Eosinophil-dependent cytotoxicity in rat schistosomiasis. Involvement of IgG2a antibody and role of mast cells. *Eur J Immunol* 1978; 8:127.
25. Thanavala YM, Bond A, Tedder R, Hay FC, Roitt IM: Monoclonal "internal image" anti-idiotypic antibodies of hepatitis B surface antigen. *Immunology* 1985; 55:197.
26. Verwaerde C, Joseph M, Capron M, et al: Functional properties of rat monoclonal IgE antibody specific for *Schistosoma mansoni*. *J Immunol* 1987; 138:4441.
27. Auriault C, Damonneville M, Verwaerde C, et al: Rat IgE directed against schistosoma-released products is cytotoxic for *Schistosoma mansoni* schistosomula in vitro. *Eur J Immunol* 1984; 14:132.
28. Damonneville M, Auriault C, Verwaerde C, Delanoye A, Pierce R, Capron A: Protection against experimental *Schistosoma mansoni* schistosomiasis achieved by immunization with schistosomula-released products antigens (SRP-A): Role of IgE antibodies. *Clin Exp Immunol* 1986; 65:244.
29. Roussel-Velge F, Verwaerde C, Grzych JM, Auriault C, Capron A: Protective effects of antiantiidiotypic IgE antibodies obtained from an IgE monoclonal antibody specific for a 26 kDa *Schistosoma mansoni* antigen. *J Immunol* 1989; 142(7):2527.
30. Joseph M, Auriault C, Capron A, Vorng H, Viens P: A role of platelets in IgG-dependent killing of schistosomes. *Nature* 1983; 303:810.
31. Bona C, Pernis B: Idiotypic networks. In: Paul WE, ed: *Fundamental Immunology*. New York: Raven Press; 1984: p 577.
32. Morrison SL, Wins LA, Oi VT: Immunoglobulin gene expression in transformed lymphoid cells in monoclonal antibodies. *Biol Clin Appl (Florence)* 1984, Abstract pL2.
33. Stein K, Soderstom T: Neonatal administration of idiotype or anti-idiotype primes for protection against *E. coli* K13 infection in mice. *J Exp Med* 1984; 160:1001.
34. Potocnjak P, Zavala F, Nussenzweig R, Nussenzweig V: Inhibition of idiotype-antiidiotype interaction for detection of a parasite antigen: A new immunoassay. *Science* 1982; 215:1637.
35. Khalife J, Capron M, Capron A, et al: Immunity in human *Schistosomiasis mansoni*. Regulation of protective immune mechanisms by IgM blocking antibodies. *J Exp Med* 1986; 164:1626.

CHAPTER 10

Idiotype Vaccines by Antibody Engineering: Structural and Functional Considerations

Maurizio Zanetti, Rosario Billetta, and Maurizio Sollazzo

"Le macchine sono effetto dell'arte, che e' scimmia della natura, e di essa riproducono non le forme ma la stessa operazione."

(Umberto Eco, *Il Nome della Rosa*)

Introduction

The idea of utilizing antibodies as noninfectious vaccines originates from the concept of idiotypes as imperfect chemical copies of antigens (1, 2) and is based on the property that antibodies can be immunogenic (3). This has been demonstrated in heterologous (4), homologous (3), syngeneic (5), and autologous (6) systems. Historically, the existence of immunoglobulin molecules related to antigenic determinants, not by fitting them but rather by resembling them, was first hypothesized by Lindemann (1). Borrowing his words there exist two different varitypes* related to one particular antigenic determinant: a varitype leading to a combining site matching that determinant, and which is conventionally called antibody; another varitype leading to an idiotypic site cross-reacting with that determinant, and for which we propose the name of homobody. Accordingly, antigenic determinants are projected into the world of immunoglobulin molecules in two manners: as negative images, these define antibodies; as positive images, these define homobodies.

Jerne subsequently renamed homobodies internal image antibodies and integrated them into the network theory of the immune system (2, 7). This concept tacitly admitted that idiotypes are shared by many antigens not belonging to the immune system and that, in a general sense, idiotypes are innate representations of a possibly infinite spectrum of epitopes in nature.

A scheme of the idea that emerged from Jerne's original intuition is shown in Figure 10.1 which illustrates two main types of three-way relationship

---

*Two immunoglobulin molecules are of the same varitype if the same amino acid residue is found at each variable position of the corresponding polypeptide chain.

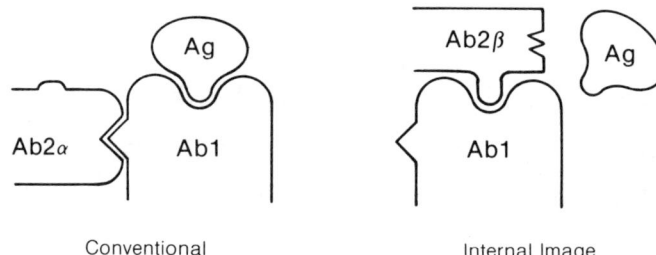

Fig. 10.1. Schematic representation of conventional (Ab2α) and internal image, antigen-mimicking, anti-Id (Ab2β) antibodies. The three-way interaction between antigen (Ag), idiotypic antibody (Ab1), and anti-Id (Ab2) antibody is illustrated. Antigen mimicry by the Ab2 molecule allows its recognition by the same receptor for the corresponding nominal antigen. In this case, the antigen receptor is the binding site of the Ab1 molecule. (From Ref. 8, with permission.)

between antigen, the first set of antibodies (Ab1), and anti-idiotypic (anti-Id) antibody (Ab2). Accordingly, Ab2 molecules in some cases recognize idiotypes only (Ab2α), in others they mimic the epitope (Ab2β) against which their complementary idiotype was directed. In other terms, Ab2 molecules can substitute for antigen in its interaction with physiological receptors and ligands to become a functional surrogate antigen.

Aside from theoretical considerations, numerous remarkable accomplishments have been made to confirm the validity of the original prediction. Idiotypes have been shown to functionally mimic exogenous, allo-, and self-antigens, tumors, hormones, and lymphokines (for review, see 8–12, 13). The many experimental demonstrations will be reviewed no doubt by other contributors to this volume. Therefore, we will not digress on them. Interestingly, their function is not restricted to protein antigens but also to carbohydrates (14–16). Although the molecular basis for this phenomenon is more difficult to understand, the general implication is that idiotypes can be used for the development of vaccines in a large variety of situations, that is, idiotypes can be considered "universal" antigens.

Structural Considerations

A rational approach to the design and development of idiotypic vaccines by genetic engineering techniques, which is the subject of this chapter, imposes some general considerations on the structure of antibodies and their idiotype. Antibodies are tetrameric molecules consisting of two identical heavy (H) chains joined with two identical light (L) chains by disulfide bonds and held together by inter-chain disulfide linkages. The two functional elements, the Fab fragment—which includes V_H and V_L and the COOH-terminal portion of

C_H1—and the Fc fragment—which consists of the remaining C_H domains—are the constitutive moieties of antibody molecules.

The genetic organization and the molecular events leading to the expression of immunoglobulin genes have been elucidated through the studies of the past decade to a great extent. The synthesis of immunoglobulin molecules occurs through a cascade of tissue-specific genetic events (for review, see 17). The H chain V region is encoded by an exon assembled upstream from the constant (C) region-coding sequence formed by the joining of three different gene elements: variable (V_H), diversity (D), and joining (J_H) segments (18). The V region of each L chain isotype (kappa and lambda) is encoded by an upstream exon also assembled from variable (V_L) and joining (J_L) segments that are joined directly within each other. The juxtaposition of these genetic elements during B cell differentiation contributes to create the diversity of the V-region repertoire, that is, antigen binding and idiotypy. This is further augmented by somatic mutation events and template-free nucleotide addition at the V–D–J joining (19).

The Fab fragment of immunoglobulin is responsible for antigen binding and idiotype expression, whereas, the Fc fragment determines the isotype and several effector functions, for example, complement fixation and binding to Fc receptors. In 1970 Wu and Kabat (20) pointed out the existence of discrete areas of variability within antibody V regions and introduced the notion of hypervariable (HV) or complementarity-determining regions (CDR). Indeed, most idiotypes mapped to date are localized in the CDRs (see below). With the resolution of the first X-ray crystal (21, 22), the three-dimensional structure of antibody and the topology of sequence conservation has become clearer. Thus, framework regions (FR) are organized as β-strands interconnected with HV loops. The FRs are arranged in a β-sheet sandwich (the "immunoglobulin fold") filled with immunoglobulin-packed hydrophobic side chains and an invariant disulfide bond linking the two sheets. The constraints on the side chains to preserve the β-folding are sufficiently stringent to explain the sequence conservation of these regions. However, HV regions can drift in sequence and conformation, with little or no effect on the structure of the framework regions.

The characterization of idiotypes has proven difficult in most systems and with rare exceptions the structural/functional correlate of idiotypy remains an enigma. There are far more idiotypic systems that have been and are being studied than cases in which the structural correlate of idiotypy is known. Part of the problem originates from the fact that, idiotypes are antigenic determinants defined by antibodies (anti-idiotypes) that are not ideal probes for the characterization of protein surface determinants. Moreover, protein antigenic determinant idiotypes can be divided into two categories, continuous and discontinuous (23, 24, 25). The first are composed of amino acid residues contiguous in the primary structure, "linear determinants". The second consist of amino acid residues discontinuous in the protein sequence that originate from residues contributed by different regions of the molecule

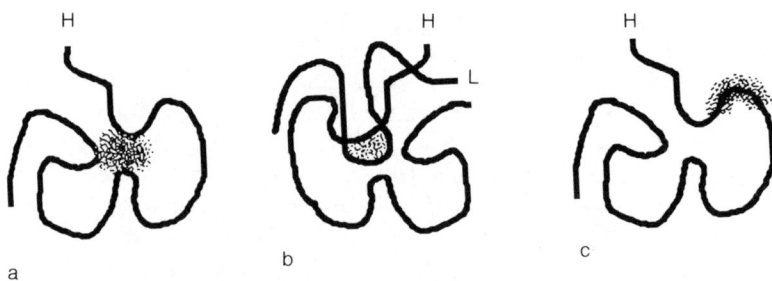

Fig. 10.2. A simplified schema of conformational idiotypes determined by discontinuous amino acids located (a) on one chain only or (b) on the two H and L polypeptide chains (type 1), and (c) a linear idiotype determined by contiguous amino acids on the same polypeptide chain (type 2).

brought together by the folding of the polypeptide to its native structure, "conformational or combinatorial determinants." These concepts are schematically represented in Figure 10.2, where three possibilities are considered. Thus, idiotypes can result from the interaction among amino acid residues within (a) one polypeptide chain only, (b) two polypeptide chains (conformational determinants), or (c) by adjacent residues on one chain only, a linear determinant.

Serologically defined idiotypes have been localized on the H (26–33), the L (34, 35), or on both H and L chains (36–39). Moreover, examples have been described in which the idiotype is contributed predominantly by one chain with an additive effect by the other chain (40). As mentioned above, the instances in which idiotype in one chain has been mapped to a precise region are limited. Although the cumulative data are so far scarce, it appears that residues in the CDR3 of the H chain (26, 31, 32, 41, 42) contribute most often to idiotype expression, albeit the CDR2 of both the H and L chains have also been implicated (35, 43, 44). This suggests that the loop constituting CDR3, which by crystallographic studies appears to be protruding from the supporting β-sheet framework, is both more accessible to solvent, that is, hydrophilic and is more mobile than other HV regions. This could explain why the CDR3 is a preferred site of antigenic determinants.

Several factors may influence idiotype expression on antibody V regions. Among these are the size, the shape, the presence of contact residues, the flexibility of the polypeptide chain, and the relative mobility of one polypeptide chain with respect to the other. In general idiotypes as antigenic determinants of immunoglobulin molecules may be seen as convex sites with large polar side chains that provide contact bonds of the ionic, hydrogen, and van der Waals types for receptors, ligands, and anti-Id antibodies. It cannot be excluded that concave idiotypes may also exist, but it is more difficult to envision the way they can interact within the surrounding molecular environment. All these factors, alone or in combinations, determine to what extent idiotypes as protein surface determinants are subject to molecular recognition and candidates for the immune function.

From the above discussion, it is apparent that antigenic determinants of immunoglobulins do not differ from those of other protein antigens, although they may be subject to additional constraints of a physicochemical nature imposed by the three-dimensional folding of the molecule. Moreover, on the basis of the available experimental and theoretical knowledge it is possible to predict where foreign epitopes might be engineered so that both the immunochemical and antigenic characteristics are maintained (see below).

Conventional Internal Image Idiotypes

The practical application of idiotypes as the internal image of antigens has been sought and successfully put into practice in many systems. Here, rather than listing them, we would like to discuss practical issues relating to their production, since the same considerations prompted us to use protein engineering techniques as an alternative approach to the preparation of internal image idiotypic antibodies. Conventionally these are borne on anti-idiotype antibodies generated by active immunization with the idiotype or the antigen and can be either monoclonal or polyclonal. (It is obvious, however, that because of the requirement for homogeneity and the cost-effectiveness of the manufacturing process, only monoclonal idiotypes will be developed.) Their clonal and molecular relationship is schematically illustrated in Figure 10.3. Although the principle is simple, that is, one needs to identify and isolate the appropriate clone, it is common experience that Ab2

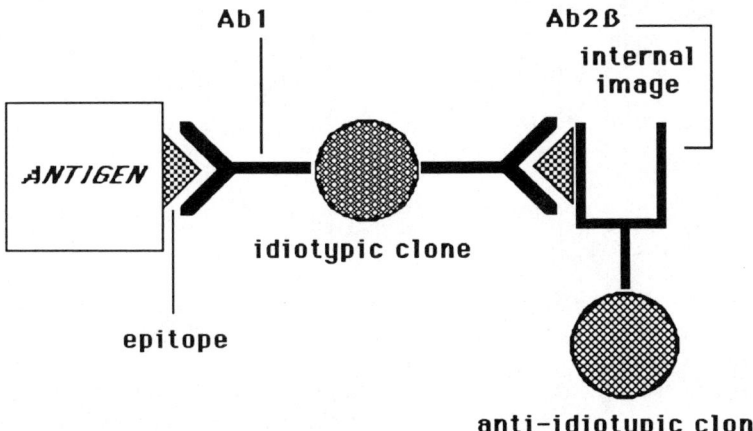

Fig. 10.3. Schematic view of the clonal relation between the epitope on the antigen, the idiotypic clone secreting Ab1, and the anti-Id clone secreting Ab2 carrying the molecular mimic of the original epitope. The Ab2 is recognized by the same receptor on Ab1 as the epitope on the nominal antigen.

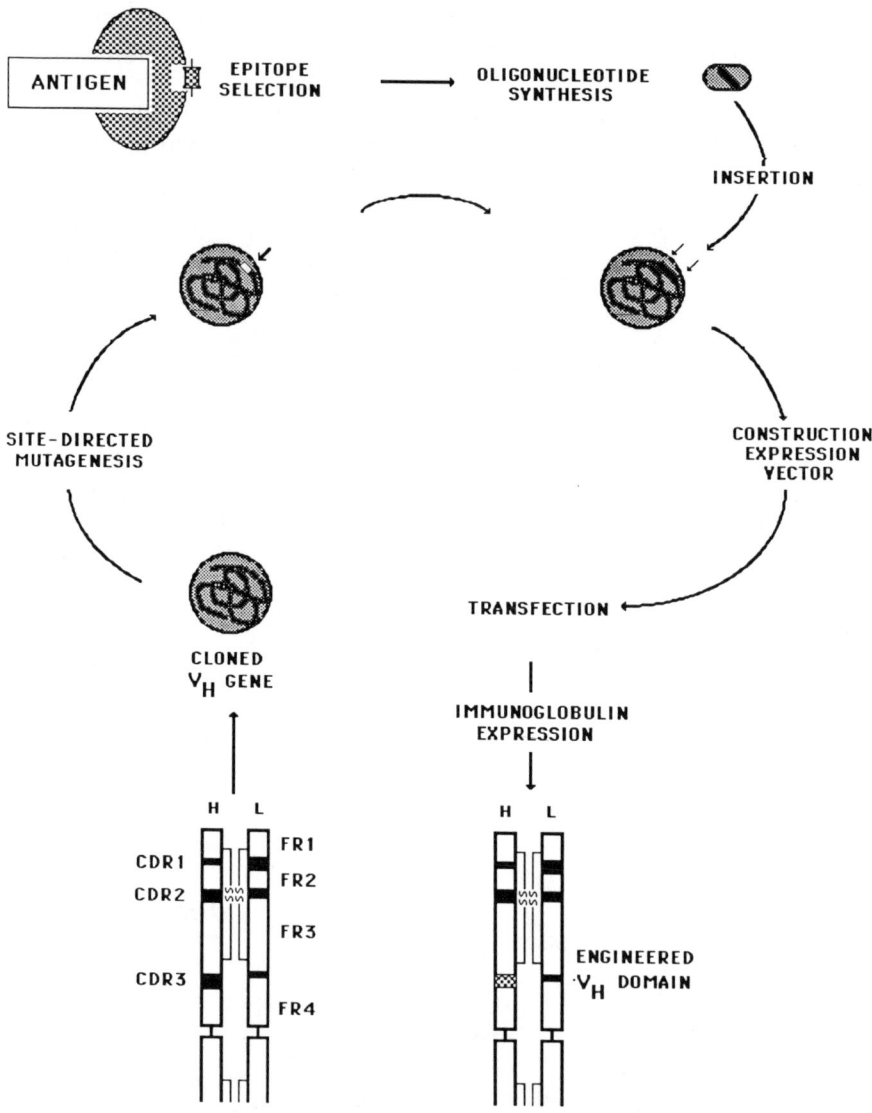

Fig. 10.4. Schematic diagram of idiotype engineering. A plasmid encoding a human or murine V_H genomic rearrangement is modified by site-directed mutagenesis to introduce (if absent) a unique restriction site into the CDR3 sequence. Subsequently, a double-stranded synthetic DNA fragment encoding the epitope of interest is inserted, in frame, into the novel restriction site and the construction verified by dideoxy sequencing (52). A restriction fragment encoding the engineered V_H gene is gel-purified and subcloned into an expression vector upstream from a human constant region gene. The H chain expression vector is electroporated into the murine J558L cell line. This host cell line is an H chain-derivative variant of myeloma J558 that carries the rearrangement for a lambda-1 L chain (53). Transfectoma cells are cultured, subcloned,

molecules are difficult to obtain. In one study in which antigen was used as the immunizing agent, internal image anti-Id antibodies were found to constitute a sizable fraction of all the clones grown in culture (45). In general, however, the clonal frequency of antigen-mimicking antibodies is low, a fact that renders the screening process long and tedious. An additional concern is that Ab2 molecules need to be evaluated in vivo for immunogenicity, a prerequisite for antigen specificity and potential therapeutic applicability. Finally even if the two previous points are met, no structural information is available that may help our understanding of the molecular basis for antigen mimicry and eliminate legitimate concerns linked with potential cross-reactivity with other antigens. Only a few studies have been reported in which a sequence homology of one HV loop with the nominal antigen was found (46–48). By and large, how idiotypes mimic antigens and therefore what the structural correlate for the induction of immunity of predetermined specificity using antibodies may be remains poorly understood.

Recently, however, it was demonstrated that the CDRs of an antibody can replace the CDRs of another Ig molecule and retain the antigen specificity of the original antibody (49–51). Using this principle, we recently began experiments to verify whether, by using protein engineering techniques, idiotypes could be built de novo into the CDRs of an immunoglobulin molecule with two objectives in mind. One, an a priori knowledge of the chemical specificity of the idiotype expressing the molecular mimic of an antigen; the other, a more stringent control of the process involved in the preparation of such an immunoglobulin molecule. In a general sense we designed a system in which the time factor and immunochemical characteristics of the final idiotypic vaccine could be optimized and controlled. A schema of the various steps involved in such an approach is shown in Figure 10.4. In the next section we will briefly describe our first attempt at engineering an idiotype.

Engineered Idiotypes

Mimicry of epitopes on antigens by immunoglobulin idiotypes can be of two types: The first is conformational, in which noncontiguous amino acids come in contact in the three-dimensional folding of the molecule and constitute a

◄――――――――――――――――――――――――――――――――

(Fig. 10.4. *continued*) and screened for the secretion of the engineered immunoglobulin molecule using a sandwich enzyme-linked immunosorbent assay (ELISA) with goat antihuman antibodies. Clones producing 10 to 30 μg/ml of protein/10^6 cells are selected and expanded, and the chimeric protein is purified by means of affinity chromatography on a 6% agarose-(recombinant)protein A-column (Repligen, Cambridge, MA). The purified immunoglobulin molecule is analyzed by SDS-PAGE under reducing and nonreducing conditions to determine both the degree of purity and molecular assembly. Key to abbreviations: CDR, complementarity-determining regions; FR, framework regions.

stereochemical copy of an antigen (we will refer to this as type 1); the second is linear, in which a stretch of the antibody molecule shares an identical amino acid sequence with the antigen (we will refer to this as type 2). As discussed earlier, some idiotypes localize in the CDR3, as its residues are exposed to solvent, "stick out" from the surface of the molecule (we will refer to this as type 2 antigen mimics). Based on this reasoning, we decided to engineer the CDR3 of the V_H of a murine immunoglobulin to be the faithful molecular mimic of the immunodominant epitope of an antigen unrelated to immunoglobulins.

We produced a chimeric (mouse/human) immunoglobulin molecule expressing three copies of the repetitive tetrapeptide Asn–Ala–Asn–Pro (NANP) of the malaria parasite *Plasmodium falciparum* circumsporozoite (CS) protein in the CDR3 of a murine V_H chain domain (55). The experimental design is schematically shown in Figure 10.5. The V_H gene encoding this epitope in CDR3 was engineered in vitro by site-directed mutagenesis, inserted into the appropriate expression vector, and used to transfect murine myeloma cells that provided the L chain. The engineered immunoglobulin was properly assembled and secreted. The insertion of a new epitope into the CDR3 of the V_H did not appreciably alter its folding and ability to associate with the L chain.

By using a monoclonal antibody generated against the parasite, and specific for $(NANP)_n$, we demonstrated that the engineered molecule expressed the $(NANP)_3$ epitope in a stereochemical configuration immunologically similar to the native CS protein. This antibody specifically recognized the $(NANP)_3$ epitope within the CDR3. An immunoglobulin molecule expressing the wild-type (WT) V_H gene and lacking this antigenic determinant was not recognized. The localization of the $(NANP)_3$ epitope on the H chain of the engineered antibody was demonstrated by Western blot. The immunochemical characterization is summarized in Table 10.1. In vivo studies in rabbits (56) indicated that the engineered immunoglobulin carrying the $(NANP)_3$ epitope is immunogenic, since it readily induced anti-NANP antibodies. These bound, by a variety of methods, the synthetic peptide $(NANP)_3$, the engineered antibody expressing the $(NANP)_3$ epitope, and the native CS protein on the *Plasmodium falciparum* parasite by indirect immunofluorescence. Serum antibodies purified by affinity chromatography on a Sepharose-4B column coated with the $(NANP)_3$ peptide reacted with the immunoglobulin molecule expressing the engineered idiotype. The ensemble of these results established a molecular symmetry and an immunological cross-reactivity between the genetically engineered idiotype, the synthetic version of it, and the natural antigen.

Several provisional conclusions can be drawn from these studies. One is that the antigenic determinant newly created in the CDR3 behaves as if it were an original determinant of the immunoglobulin molecule. Neither the molecular environment nor the globular folding appear to have modified the immunological structure of the $(NANP)_3$ epitope inserted into the CDR3 by

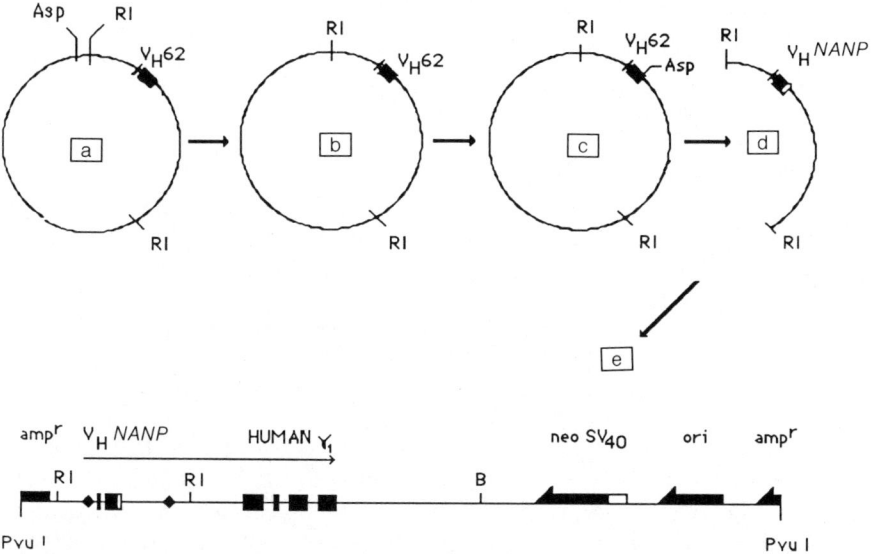

Fig. 10.5. Diagram illustrating the construction of the pNγ1 *NANP* expression vector. (a) The productively rearranged V_H gene of the hybridoma cell line 62 isolated from a size-selected lambda gt10 library and subcloned into pBlue-script (STRATAGENE, San Diego, CA) was described previously (33). (b) The restriction site Kpn I/Asp 718 of the vector polylinker region was deleted by Kpn I digestion, filled in with T4 polymerase, and ligated to yield the plasmid pH62Δk. (c) pH62Δk was used as a template for site-directed mutagenesis to introduce a unique Asp718 restriction site in CDR3 of the V_H gene. The synthetic oligonucleotide (5′CAAGAAAGG*TAC*CCTACTCTC3′), which encodes a 3 bp insertion (TAC), was annealed to the uracylated-single–stranded complementary template and elongated as described by Kunkel (54). (d) Complementary synthetic oligonucleotides,

(5′GTACCCAATGCAAACCCAAATGCAAACCCAAATGCAAACCCA3′-
5′GTACTGGGTTTGCATTTGGGTTTGCATTTGGGTTTGCATTGG3′)

encoding the (NANP)₃ peptide, and carrying Asp718 protruding ends, were annealed and subcloned into the unique Asp718 site of pH62k vector. The construction was verified by sequence analysis using a 15^{mer} primer corresponding to the 5′ end of the V_H62 gene (5′GACGTGAAGCTGGTG3′). (e) The 2.3 kb EcoR1 fragment carrying the engineered V_H NANP gene was subcloned upstream from the human γ1 constant region gene into the 13-kb pNγ1 vector (33). The Pvu I-linearized pNγ1 NANP construct was electroporated into J558L cells and stable-transformant selected in the presence of G418 as described (33). Key to abbreviations: Asp, Asp 718; B, Bam H1; R1, EcoR1; amp^r, ampicillin resistance; neo, neomycin (G418) resistance; SV_{40}, SV_{40} early-promoter elements; ori, pBR322 origin of replication; (♦), mouse immunoglobulin, heavy chain, transcriptional regulatory elements.

Table 10.1. Immunochemical characterization of the engineered molecule.

Recombinant immunoglobulin	Binding of [^{125}I]-labeled monoclonal antibody specific for the NANP sequence	
	RIA	Western blot
γ1 NANP[a]	+	+ (heavy chain)
WT	−	−

[a] Recombinant chimeric mouse/human immunoglobulin engineered to express the (NANP)$_3$ epitope in the CDR$_3$ of V$_H$.
[b] Recombinant chimeric mouse/human immunoglobulin of the wild-type V$_H$ lacking the (NANP)$_3$ epitope.

genetic manipulation. The second is that as far as we could determine, the type 2 engineered idiotype acquired the characteristics of the CDR3 region where it was inserted, for example, reactions to solvent and immunogenicity. Although the findings may be strictly related to the choice of this particular antigenic determinant, a generalization would be that type 2 idiotypes can be engineered in vitro by inserting discrete antigenic determinants of "foreign" antigens into the CDRs of a host immunoglobulin V region domain. It remains to be established how large epitopes can be engineered into immunoglobulins CDRs, whether such a manipulation can be done in all CDRs, and what the limitations of this approach may be.

Theoretical Aspects and Practical Applications of Engineered Idiotype Vaccines

The results obtained indicate that idiotypes genetically engineered into a host immunoglobulin V region, idiotypes à la carte, can be expected to have the immunochemical characteristics of their original native protein and be exploited for the induction of immunity of predetermined specificity. This new approach offers a series of potentially interesting advantages. Multiple idiotypes can be engineered in the same molecule to mimic the same or different epitopes of the antigen against which immunity is sought. Theoretically this should be possible, since the V_H and V_L polypeptide chains can tolerate several changes in the CDRs amino acid sequence while maintaining the unaltered immunoglobulin total (for a review, see reference 57). This principle has already been demonstrated for the CDRs of human immunoglobulins that can be substituted with the CDRs of murine antibodies, "humanized antibodies" (49–51). It will be important to determine what the role of residues between the strands and the framework regions might be in determining a proper spatial orientation of the engineered idiotype, that is, its immunogenic property.

Type 2 idiotypes can be engineered to elicit B cell or T cell immunity preferentially. In the latter case one may choose to engineer epitopes for the activation of helper cells or cytotoxic T cells, depending upon the type of pathogen and the immunity required for protection. Obviously, an a priori definition of these antigenic determinants on the native protein is necessary. This information can be obtained by using established algorithms and criteria of prediction based on epidemiological studies. It is important that the engineered idiotypes are (a) immunodominant in the largest possible assortment of major histocompatibility complex (MHC) haplotypes and (b) of demonstrable biological relevance in protection from disease. The above assumptions do not take into consideration, however, the way idiotypes are presented to immunocompetent cells. In fact, it is not yet clear whether immunoglobulins are processed and presented in association with MHC molecules as conventional nonimmunoglobulin antigens. Given these unknowns on the physiology of idiotypes as antigens, we predict that by the approach described here one can render a B cell epitope T independent. As far as the triggering of cytotoxic T cells (CTL) is concerned, recent evidence indicates that they can be induced by soluble proteins (58, 59). The success of engineered idiotypes in this function may ultimately depend upon the knowledge that will be acquired using the mechanism(s) whereby antibody V regions are presented to CTL. In particular we will need to know how immunoglobulins are enzymatically cleaved in the intracellular compartment and associate with MHC class I antigens, in case this were required. Experiments are in progress to test the validity of these hypotheses.

An obvious limitation of this approach is that only protein epitopes will be amenable to conversion into type 2 idiotypes by protein engineering. Carbohydrates, which are important for protection against numerous bacteria and parasites, require different strategies. There exists, however, the possibility that the shape of an oligosaccharide may be mimicked by oligopeptides, though this process requires the use of sophisticated chemical synthesis procedures at this stage (60). It cannot be excluded that, in the future, by the process of optimization of chemical and protein engineering techniques, one may be able to manipulate the immune system using engineered idiotypes that mimic carbohydrate epitopes. This will be of considerable significance in the ability to involve T cells directly in the induction of immunity.

The prospect for the application of engineered idiotype vaccines is promising and potentially applicable to a wide range of diseases, particularly if all of the above considerations and predictions of structural and immunological nature prove to be correct. Their use could be either for prophylaxis or immunotherapy against parasitic, viral, and bacterial pathogens for which there is hope that pathology and disease can be prevented or reversed by appropriate immune intervention. In a not-so-distant future, such an approach may also be applicable to the control of autoimmune diseases. Before they can be used clinically, recombinant antibody molecules bearing engineered idiotypes as described herein will be useful as probes for the immunogenic function of protein antigens and the molecular basis of the structure–function relation of the immunogenicity of immunoglobulin V regions.

Acknowledgments. This work was supported in part by grants from the Council for Tobacco Research #2124R1; NIH #AI 23871; and The Immune Response Corporation. M.Z. is a scholar of the Leukemia Society of America, Inc.; The authors are grateful to Mrs. Debbie Lundemo for preparation of this manuscript.

References

1. Lindenmann J: Speculations on idiotypes and homobodies. *Ann Immunol* 1973; 124:171–184.
2. Jerne NK: Towards a network theory of the immune system. *Ann Immunol* 1974; 125:373–389.
3. Oudin J, Michel M: Une Nouvelle forme d'allotypie des globulines du serum de lapin apparemment liee a la fonction et a la specificite anticorps. *CR Acad Sci* 1963; 257:805–808.
4. Kuettner MG, Wang AL, Nisonoff A: Quantitative investigations of idiotypic antibodies. VI. Idiotypic specificity as a potential genetic marker for the variable regions of mouse immunoglobulin polypeptide chains. *J Exp Med* 1972; 135:579–595.
5. Sakato N, Eisen HN: Antibodies to idiotypes of isologous immunoglobulins. *J Exp Med* 1975; 141:1411–1426.
6. Rodkey LS: Studies of idiotypic antibodies. Production and characterization of autoantiidiotypic antisera. *J Exp Med* 1974; 139:712–720.
7. Jerne NK, Roland J, Cazenave PA: Recurrent idiotopes and internal images. *EMBO J* 1982; 1:243–247.
8. Zanetti M, Sercarz E, Salk J: Immunology of new generation vaccines. *Immunol Today* 1987; 8:18–25.
9. Bona C, Moran T: Idiotype vaccines. *Ann Immunol* 1985; 136C:299–311.
10. Zanetti M, Katz DH: Self recognition, autoimmunity and internal images. In: Koprowski H, Melchers F, eds: *Current Topics in Microbiology and Immunology.* Heidelberg: Springer-Verlag; 1985:pp 111–126.
11. Stevenson GT, Stevenson FK: Treatment of lymphoid tumors with anti-idiotype antibodies. *Springer Semin Immunopathol* 1983; 6:99–115.
12. Strosberg AD: Auto-idiotype and anti-hormone receptor antibodies. *Springer Semin Immunopathol* 1983; 6:67–78.
13. Zuberi RI, Katz DH, Zanetti M: Production of IL-1-mimicking anti-idiotypic antibodies in rabbits in response to IL-1 immunization. *J Autoimmunity* 1988; 1:31–46.
14. Reagan KJ, Wunner WH, Wiktor TJ, Koprowski H: Anti-idiotypic antibodies induce neutralizing antibodies to rabies virus glycoprotein. *J Virol* 1983; 48:660–666.
15. Stein KE, Soderstrom T: Neonatal administration of idiotype or antiidiotype primes for protection against *Escherichia coli* K13 infection in mice. *J Exp Med* 1984; 160:1001–1011.

16. Sacks DL, Kirchhoff LV, Hieny S, Sher A: Molecular mimicry of a carbohydrate epitope on a major surface glycoprotein of *Trypanosoma cruzi* by using anti-idiotypic antibodies. *J Immunol* 1985; 135:4155–4159.
17. Alt FW, Blackwell TK, Yancopoulos GD: Development of the primary antibody repertoire. *Science* 1987; 238:1079–1087.
18. Early P, Huang H, Davis M, Calame K, Hood L: An immunoglobulin heavy chain variable region gene is generated from three segments of DNA: V_H, D, and J_H. *Cell* 1980; 19:981–992.
19. Alt FW, Baltimore D: Joining of immunoglobulin in heavy chain gene segments: Implications from a chromosome with evidence of three D-J_H fusions. *Proc Natl Acad Sci USA* 1982; 79:4118–4122.
20. Wu TT, Kabat EA: An analysis of the sequences of the variable regions of Bence-Jones proteins and myeloma light chains and their implications for antibody complementarity. *J Exp Med* 1970; 132:211–250.
21. Poljak RJ, Amzel LM, Avey HP, Chen BL, Phizackerly RP, Saul F: Three-dimensional structure of the Fab' fragment of a human immunoglobulin at 2.8 Å resolution. *Proc Natl Acad Sci USA* 1973; 70:3305–3310.
22. Schiffer M, Girling RL, Ely KR, Edmundson AB: Structure of a λ-type Bence-Jones protein at 3.5 Å resolution. *Biochemistry* 1973; 12:4620–4631.
23. Novotny J, Handschumacher M, Haber E, et al: Antigenic determinants in proteins coincide with surface regions accessible to large probes (antibody domains). *Proc Natl Acad Sci USA* 1986; 83:226–230.
24. Barlow DJ, Edwards MS, Thornton JM: Continuous and discontinuous protein antigenic determinants. *Nature* 1986; 322:747–748.
25. Thornton JH, Edwards MS, Taylor WR, Barlow DJ: Location of "continuous" antigenic determinants in the protruding regions of proteins. *EMBO J* 1986; 5:409–413.
26. Rudikoff S, Pawlita M, Pumphrey J, Mushinski E, Potter M: Galactin-binding antibodies. Diversity and structure of idiotypes. *J Exp Med* 1983; 158:1385–1400.
27. Stanislawski M, Pene J: A common V_H marker relating BALB/c alpha 1-3 dextran-binding and A/J P-azophenylarsonate-binding antibody families. *J Immunol* 1983; 130:2434–2441.
28. Zeldis JB, Konigsberg WH, Richards FF, Rosenstein RW: The location and expression of idiotypic determinants in the immunoglobulin variable region. II. Chain location of variable region determinants. *Mol Immunol* 1979; 16:371–378.
29. Kobzik L, Brown MC, Cooper AG: Demonstration of an idiotypic antigen on a monoclonal cold agglutinin and on its isolated heavy and light chains. *Proc Natl Acad Sci USA* 1976; 73:1702–1706.
30. Zanetti M, Liu F-T, Rogers J, Katz DH: Heavy and light chains of a mouse monoclonal autoantibody express the same idiotype. *J Immunol* 1985; 135:1245–1251.
31. Borden P, Kabat EA: The specificities of polyclonal and monoclonal anti-idiotypes to anti-α(1-6)dextrans; possible correlations of idiotype with amino acid sequence. *Mol Immunol* 1988; 25:251–262.
32. Gridley T, Margolies MN, Gefter ML: The association of various D elements with a single immunoglobulin V_H gene segment: Influence on the expression of a major cross-reactive idiotype. *J Immunol* 1985; 134:1236–1244.
33. Sollazzo M, Hasemann CA, Meek KD, Glotz D, Capra JD, Zanetti M: Molecular characterization of the V_H region of murine autoantibodies from neonatal and adult BALB/c mice. *Eur J Immunol* 1989; 19:453–457.

34. Vrana M, Rudikoff S, Potter M: The structural basis of a hapten-inhibitable k-chain idiotype. *J Immunol* 1979; 122:1905–1910.
35. Chen PP, Fong S, Normansell D et al: Delineation of a cross-reactive idiotype on human autoantibodies with antibody against a synthetic peptide. *J Exp Med* 1984; 159:1502–1511.
36. Suzan M, Boyer C, Schiff C, Trucy J, Milili M, De Preval C: Monoclonal antibodies reveal three types of idiotypic determinants on MOPC 173: H-L-specific private and public idiotopes, and V_H-specific public idiotope(s). *Mol Immunol* 1982; 19:1051–1062.
37. Lee MS, Horng WJ, Gilman-Sachs A, Dray S: The common idiotypic specificity of anti-a2 allotype antibodies: Contribution of H and L chains to idiotype expression. *Mol Immunol* 1983; 20:557–561.
38. Kranz DM, Voss EW: Idiotypic analysis of monoclonal anti-fluorescyl antibodies: Localization and characterization of idiotypic determinants. *Mol Immunol* 1983; 20:1301–1322.
39. Robbins PF, Rosen EM, Haba S, Nisonoff A: Relationship of V_H and V_L genes encoding three idiotypic families of anti-azobenzenearsonate antibodies. *Proc Natl Acad Sci USA* 1986; 83:1050–1054.
40. Nusair T, Baumal R, Rosenstein R, Jorgensen T, Marks A: Characterization of idiotopes on MOPC 315 IgA using monoclonal antiidiotypic antibodies. *Mol Immunol* 1983; 20:537–547.
41. Berek C: The D segment defines the T15 idiotype: The immune response of A/J mice to *Pneumococcus pneumoniae*. *Eur J Immunol* 1984; 14:1043–1048.
42. Sollazzo M, Castiglia D, Billetta R, Tramontano A, Zanetti M: Structural definition by antibody engineering of an idiotypic determinant *Prot Engng* 1990; 3:531–539.
43. Clevinger B, Schilling J, Hood L, Davie JM: Structural correlates of cross-reactive and individual idiotypic determinants on murine antibodies to (1-3) dextran. *J Exp Med* 1980; 151:1059–1070.
44. Cleary ML, Meeker TC, Levy S, et al: Clustering of extensive somatic mutations in the immunoglobulin heavy chain variable region gene of a human B cell lymphoma. *Cell* 1986; 44:97–106.
45. Cleveland WL, Wassermann NH, Sarangarajan R, Penn AS, Erlanger BF: Monoclonal antibodies to the acetylcholine receptor by a normally functioning auto-anti-idiotypic mechanism. *Nature* 1983; 305:56–57.
46. Mazza G, Ollier P, Somme G, et al: A structural basis for the internal image in the idiotypic network: Antibodies against synthetic Ab2-D regions cross-react with the original antigen. *Ann Inst Pasteur/Immunol* 1985; 136D:259–269.
47. Bruck C, Co MS, Slaoui M, et al: Nucleic acid sequence of an internal image-bearing monoclonal anti-idiotype and its comparison to the sequence of the external antigen. *Proc Natl Acad Sci USA* 1986; 83:6578–6582.
48. Van Cleave VH, Naeve CW, Metzger DW: Do antibodies recognize amino acid side chains of protein antigens independently of the carbon backbone? *J Exp Med* 1988; 167:1841–1848.
49. Jones PT, Dear PH, Foote J, Neuberger MS, Winter G: Replacing the complementarity-determining regions in a human antibody with those from a mouse. *Nature* 1986; 321:522–525.
50. Riechmann L, Clark M, Waldmann H, Winter G: Reshaping human antibodies for therapy. *Nature* 1988; 332:323–327.

51. Verhoeyen M, Milstein C, Winter G: Reshaping human antibodies: Grafting an antilysozyme activity. *Science* 1988; 239:1534–1536.
52. Sanger F, Nicklen S, Coulson AR: DNA sequencing with chain-terminating inhibitors. *Proc Natl Acad Sci USA* 1977; 74:5463–5467.
53. Morrison SL: Transfectomas provide novel chimeric antibodies. *Science* 1985; 229:1202–1207.
54. Kunkel TA, Roberts JD, Zakour RA: Rapid and efficient site-specific mutagenesis without phenotypic selection. *Methods Enzymol* 1987; 367:381.
55. Sollazzo M, Billetta R, Zanetti M: Expression of an exogenous peptide epitope genetically-engineered in the variable domain of an immunoglobulin. Implications for antibody and peptide folding. Prot Engng 1990; in press.
56. Billetta R, Hollingdale MR, Zanetti M: Antibody engineered to carry the (NANP)$_3$ epitope in the variable region induces immunity to malaria. Submitted for publication.
57. Alzari PM, Lascombe M-B, Poljak RJ: Three-dimensional structure of antibodies. *Ann Rev Immunol* 1988; 6:555–580.
58. Yewdell JW, Bennink JR, Hosaka Y: Cells process exogenous proteins for recognition by cytotoxic T lymphocytes. *Science* 1988; 239:637–640.
59. Moore MW, Carbone FR, Bevan MJ: Introduction of soluble protein into the class I pathway of antigen processing and presentation. *Cell* 1988; 54:777–785.
60. Goodman M, Coddington J, Mierke DF, Fuller WD: A model for the sweet taste of stereoisomeric retro-inverso and dipeptide amides. *J Am Chem Soc* 1987; 109:4712–4714.

Index

Ab2α, 8, 10–18, 31, 32, 35, 50, 74, 84, 124
Ab2β, 8, 10–18, 31–37, 42, 50, 55, 74, 83, 84, 111, 118, 124, 127
Ab2ε, 9, 74
Ab2γ, 74
Agretope, 5
AIDS, 1, 77, 79, 82, 96, 97, 100
Allotypes, 3, 5, 8, 11, 62, 63, 108
Arsonate, 4
Autoimmunity, 76

B cell repertoire, 22

CD4, 18, 43, 76–81, 92, 97–103
CD8, 18
Cross-reactive idiotype, see Recurrent idiotypes
CTL, 60, 64, 65, 75, 133
Cyclophosphamide, 68, 69

Dendritic cells, 5
DNA antigen, 27

Epibody, see Ab2ε

Fructosan, 15

HBs Ag, see Hepatitis B virus
Hemophilus influenzae, 73, 85

Hepatitis B virus, 16, 43, 73, 82–84
HIV, 43, 76–82, 85, 96–103
Homobody, see Ab2β

Individual idiotypes, see Private idiotype
Influenza virus
 hemagglutinin, 12
 neuraminidase, 12
Insulin, 15, 23
Internal image, 9, 15, 16, 35, 42, 43, 46, 51, 57, 59, 64, 79, 80, 93, 108, 109, 111, 119, 123, 127, 129
Isotype, 115–118, 125

Leishmania major, 109
Levan, 15, 32
Listeria monocytogenis, 15
Lymphokines
 IL2, 15
 IL4, 11

Melanoma, 14, 59
MHC
 class II, 18, 77
 haplotype, 133
 HLA, 5, 18
 MHC-restriction, 3, 5, 10, 11, 15, 17, 18, 62, 63, 65
Mycobacterium tuberculosis, 18

Neisseria meningitidis, 73
Network theory, 1, 8, 42, 50, 55, 62, 83, 118, 123
NP, 34, 35
Nude mice, 3, 17

Paratope-induced antibodies, *see* Ab2β
Plasmodium berghei, 120
Poliovirus, 17
Polyreactive antibodies, 23
Private idiotype, 62, 80
Processing, 3, 5, 17, 62, 63, 64, 65

Recurrent idiotypes
  anti-*Candida albicans*, 31
  anti-*coxsackievirus*, 75
  anti-cytomegalovirus, 75
  anti-*Escherichia coli*, 33
  anti-*Fuc α(1→3) Gal*, 36
  anti-*Haemophilus influenzae*, 31, 75
  anti-hepatitis B virus, 75, 82–84
  anti-herpes simplex, 75
  anti-HIV, 75, 77–82, 85, 99–103
  anti-*Leishmania major*, 109
  anti-lymphoma L1210, 60–69
  anti-MCA-1490 sarcoma cells, 56
  anti-meningococcal, 32–33
  anti-*Neisseria meningitidis*, 35
  anti-Newcastle disease virus, 75
  anti-poliovirus, 75
  anti-rabies virus, 75
  anti-reovirus, 75, 92–96
  anti-*Schistosoma mansoni*, 110–119
  anti-sendai virus, 75
  anti-streptococcal, 31, 35
  anti-SV40 transformed cells, 56
  anti-TADAp97, 57
  anti-TADAp175, 57
  anti-T-ALL, 57–59
  anti-*Trypanosoma cruzi*, 10, 34, 109
  anti-*Trypanosoma rhodesiense*, 108, 109
  A48, 15, 33
  B1-8, 34
  CRI, 2
  E109, 33
  GAT, 10
  M315, 18
  M460, 9
  MBrI, 42–51
  Reovirus, 16, 92–96

Schistosoma mansoni, 17, 110–119
Sendai virus, 14, 75, 110
Shared idiotypes, *see* Recurrent idiotypes
SIV, 80, 81
Sm, 23
SRBC, 12
Streptococcus pneumoniae, 17
Suppression
  hapten specific, 4
  idiotypic, 1

T cell acute lymphoblastic leukemia, 57–59
TCR, 8, 18
T-dependent antigens, 32, 33, 36
TH cells, 61–68, 77, 133
T-independent antigens, 11, 32
TNP, 62, 66, 67
Trypanosoma cruzi, 10, 34, 109
Trypanosoma rhodesiense, 108, 109
Tumor-associated antigens, 42–51, 55, 56

$V\beta$, 14
VH
  human, 22–29
  mouse, 29, 93–95, 119, 126, 128, 130
Vibrio cholera, 73
VL, 93–95, 126, 128